City Lights

LANDSCAPES OF THE NIGHT

George F. Thompson
Series Founder and Director

Published in cooperation with the
Center for American Places,
Santa Fe, New Mexico,
and Harrisonburg, Virginia

City Lights

Illuminating the American Night

JOHN A. JAKLE

THE JOHNS HOPKINS UNIVERSITY PRESS
Baltimore & London

For Cindy

© 2001 The Johns Hopkins University Press
All right reserved. Published 2001
Printed in the United States of America on acid-free paper
9 8 7 6 5 4 3 2 1

The Johns Hopkins University Press
2715 North Charles Street
Baltimore, Maryland 21218-4363
www.press.jhu.edu

Photographs and postcards are from the author's personal collection unless otherwise credited.

Library of Congress Cataloging-in-Publication Data
Jakle, John A.
   City lights : illuminating the American night / John A. Jakle.
      p. cm. — (Landscapes of the night)
   ISBN 0-8018-6593-X
   1. Municipal lighting—United States. 2. Municipal lighting—United States—Social aspects. I. Title. II. Series.
   TK4188 .J35 2001
   628.9'5'0973—dc21

00-011213

A catalog record for this book is available from the British Library.

Contents

Preface and Acknowledgments

I offer here a brief history of public lighting in American cities from the mid-nineteenth century through to the mid-twentieth century. My emphases are twofold. First, I trace the evolution of lamp technology from oil and gas through various kinds of electrical illumination, focusing on the different qualities of light used to illuminate the night. Second, I outline how various kinds of lamps were applied to the outdoor lighting of public space, especially streets. My purpose is to better understand how night light contributed to the modernizing of American cities, especially through its use in improved transportation. I emphasize the place-enhancing qualities of nighttime lighting, particularly as it contributed to the nation's emerging reliance on the automobile. Nighttime lighting became primarily a facilitator of auto use after dark, as public streets were made fully into traffic arteries.

Scholarly concern with the American landscape has been primarily a daytime consideration. An extensive literature focusing on American built environments has been produced by a diversity of scholars, including geographers, historians, architects, and folklorists, among others. Although students of landscape have been predominantly concerned with material culture, scholars have recently refocused on the social symbolisms inherent in landscape, especially the symbolisms of economic function and associated social class and power. Whether approached as physical feature, social symbol, or a combination of the two, landscape has been conceptualized primarily in terms of daytime use. The geography of night is just now receiving rightful attention. What makes nighttime geography possible, of course, is artificial illumination. For over a century public places, especially in cities, have been brightly lit at night, enabling the extension of many daytime activities into the after dark hours. New economic and social opportunities distinctive to the night were created through the use of lighting, as were distinctive nighttime landscapes and places.

In the nineteenth century, public spaces were illuminated to allow surveillance and thereby promote safety. Crime and other deviant behavior was thought to be diminished in well-lit places. In the twentieth century, however, night visibility became a facilitator of rapid movement, specifically, the movement of automobiles. The idea was to protect people and property not only from criminal assault but from automobile accidents. As public spaces became oriented less to pedestrians and more to motorists, many aspects of nighttime place-making were slighted, reduced in emphasis, or otherwise modified. Particularly diminished were the celebratory and informative applications of light. This book tells the story of nighttime lighting (a story of tyranny, perhaps), whereby public urban spaces at night were reconfigured to accommodate the automobile as machine.

Chapters vary in length; each is designed to crystallize ideas rather than thoroughly exhaust topics. My purpose is as much to raise questions as to answer them. My treatment of change over time differs from that of many historical analyses. As lighting technologies evolved, some were sustained over long periods, but others were abandoned, sometimes quickly, in the face of new innovation. Technological applications were introduced, some to survive to the present day and others to disappear. New lighting applications excited public comment, which fell off, often abruptly, as novelty eroded. Temporal emphasis, therefore, varies topic to topic. In general, I emphasize the first introductions and the initial applications appropriate to each topic. So also does my treatment of spatial variation differ from that of many geographical analyses. My purpose is not to map lighting applications, based on the origins and geographical spread of various lighting technologies. Rather, it is to explore artificial illumination as a mechanism for place-making. I define *places* quite simply as centers of human attention and activity.

My concern with landscape as a topic of investigation departs from usual academic preoccupation. Most students of landscape narrowly concentrate on how landscapes function. With this emphasis, they ignore (and many demean) the visual qualities of landscape. The visual elements of landscape are dismissed as mere aesthetics. Landscapes of the night would not function if sight were not made possible through artificial illumination. Seeing in the night is essential to landscape utility. This book, therefore, necessarily embraces landscape visualization as part of landscape function. Emphasis is placed on the seeing of places and on places specifically contrived to be seen.

Following an introduction, the book is organized in two sections. The chapters of Part 1 treat the rise of major lamp technologies, especially those applied to street lighting—from oil and gas lamps through electric arc, incandescent filament, and gaseous discharge lamps. In what order did various technologies appear? What were their capabilities? Their limitations? How did they serve to alter the darkness? Lighting applications are emphasized in Part 2, starting with street illumination. Just how were various lamp technologies applied in lighting public space? What were the principal strategies employed? Who was responsible? What were their motives? Also treated are the celebratory uses of nighttime lighting—the street festival, the large exposition or world's fair, the amusement park, and the floodlighting of landmarks—and the informative uses of artificial illumination, especially in advertising—the rise of New York City's Broadway as the "Great White Way," the emergence of electric sign art across the United States, and the related evolution of brightly lit retail streets, both traditional "main streets" and later-day commercial strips. In sum, I treat that period of American history in which a distinctive "geography of night" took form in the nation's cities through application of nighttime illumination.

Thanks are due all those who extended help in the researching and preparation of this book—librarians who aided in the assembly of written materials, archivists who facilitated the collection of photographs and other illustrations, and colleagues in academia, and, as well, those in private industry, who helped refine ideas. Thanks also are due several anonymous reviewers in the history of technology, in geography, and in urban studies. I am especially appreciative of George F. Thompson and his staff at the Center for American Places, and the staff of the Johns Hopkins University Press. It was George Thompson's vision for a new book series, of which this book is one of the inaugural volumes, that excited in me serious resolve to pursue a long-held interest in the nighttime lighting of American cities. Special thanks to Jane Domier, who prepared line drawings, and to Barbara Bonnell, who handled word processing.

City Lights

City Lights in
the American Imagination

To modern eyes, a nineteenth-century American city would undoubtedly appear strange at night. Could we travel back in time to the New York City of the 1880s, as in Jack Finney's science fiction classic *Time and Again,* we might find a nighttime world remotely familiar but disorienting.[1] Today's cities shine brightly at night, their public spaces fully illuminated and revealed. In contrast, we envision yesterday's cities as dominated by shadow, revealing much less to the observing eye. But just how dark—how unrevealing—were they? What were past nighttime landscapes like to those who knew them firsthand? And, more importantly, just what did improved night illumination make possible as new lighting technologies were developed and applied? What might consideration of past lighting practice teach that will be useful today in our understanding and appreciation of contemporary cities at night?

Finney used an imagined nighttime New York City to set tone, establish action, and drive plot development in his tale of time travel. Finney's hero and heroine, as novice time-travelers, are thrust abruptly into a nighttime Manhattan in pursuit of a mysterious stranger. The hero narrates the story:

> We followed, watching him pass through the yellowy circles at the base of each streetlamp, the light sliding along the silk curves of his hat. Beyond the curb Broadway lay in almost complete darkness. . . . The traffic . . . dim shapes and moving shadows visible only in bits and pieces. You'd see a fan shape of muddy spokes revolving through the swaying light of a lantern

slung from a van's axle, but the wagon itself and its driver and team would be lost in the blackness.[2]

"Across the dark street," he continues, "the windows and doorways of business houses were almost dark, their shapes silhouetted only by turned-down night-lights." "This strange dim street . . . relieved by squares, rectangles and cones of vague light whose very color was strange, had me uneasy," Finney's hero admits.[3]

Writing from the perspective of the late-twentieth century, Finney conceptualized the very heart of 1880s New York City as dim, and thus disorienting, to modern sensibilities. Writers of the day, however, considered Manhattan's nighttime Broadway brightly illuminated and even a place of visual spectacle. The editors of the *Electrical Review,* writing in 1886, remembered the first applications of gas light in shop windows along Broadway some thirty years earlier. "Owing to the brilliancy of the light pedestrians could walk by stores of the same character lit by gas without even seeing them, so attractive was the brilliant illumination further along."[4] Onlookers, they wrote, "fluttered" about lit shop windows "like moths." Lighting spread quickly once adopted in an area. "The neighboring stores must have it," the commentary continued, and "the inquiry and demand for the light spread apace until now, when, as soon as the electric light appears in one part of a locality in an American city, it spreads from store to store and from street to street."[5] Everything, of course, stands relative, exaggeration entering easily when comparisons are made across time.

CITY LIGHTS AND URBANIZATION

The lighting of American cities did not occur in isolation; it was integrally linked to all other aspects of the nation's urbanization. As the American economy evolved, artificial illumination in cities, which was a fundamentally capitalistic enterprise, served at least two principal purposes. First, it made city space functional at night, a stimulus to commercial activity and profit-making. Second, lighting, in itself, offered investment opportunity. Capitalists found the development and application of lighting technology extremely profitable—lighting the American city engaged the nation's largest financial institutions and led to creation of some its largest corporations. Lighting was tied to innovation in transportation, especially the rise of automobile use. City streets came to be lit primarily in accommodation of rapid automobiles and motor trucks.

Night lighting had profound influence on city growth and, accordingly, on evolving city morphology, long before the automobile age. Artificial illumination allowed the pedestrian-oriented city to function in the early morning and late evening, greatly standardizing the length of the workday across the seasons of the year. Work, whether at a factory, office, or store, could be better regulated, encouraging increased labor productivity and larger scales of operation. Land use became specialized and spatially segregated once a new transit-oriented city had spread out along railroad and streetcar lines and longer commutes were fostered by the lighting of public spaces used for travel. Nighttime lighting's encouragement to automobiles was part of a larger process of urban evolution, whereby movement and nighttime visualization were conjoined.

During the early decades of the nineteenth century, the typical American city was very compact, usually a pedestrian place oriented to a waterfront, either the harbor of a seaport or lake or a river levee. There commodities, including those that moved in bulk, were exchanged between land and water carriage. Buildings were tightly packed and housed a great diversity of activities, business and residential functions tending to be combined in single structures. Artisans lived where they worked, and most business activity was organized at the scale of the artisan shop. Close at hand were the more palatial dwellings of affluent businessmen, sometimes clustered on specific streets but only slightly removed from the noise and congestion of waterfront traffic. Much activity spilled out of buildings onto public streets, creating a vigorous street life. The largest cities grew largely by expanding neighborhood by neighborhood, both along and away from waterfronts. Streets and buildings, and their mix of functions, were replicated over and over again, producing largely self-contained districts, each supporting a similar range of activity. Only at the very center were the beginnings of land-use specialization clearly evident. Here one found the halls of citywide government, the most prominent churches, the most influential clubs and other social institutions, the principal hotels, and the most widely known ceremonial spaces, especially public plazas.

After 1840, railroads usurped most city waterfronts, serving as "feeder lines" strengthening urban outreach into surrounding hinterlands and vastly enlarging those hinterlands. The inward flow of agricultural and other raw material into cities was enhanced and accelerated, as was the outward flow of finished goods. Factories, with work organized increasingly around the tending of machines, grew larger and larger along the tracksides, at distances farther and farther out from city centers. There, new residential neighborhoods were constructed to house workers.

Not only did new neighborhoods evolve with degrees of specialization not previously known, but older neighborhoods also changed. Warehouses and industrial loft buildings intruded, displacing the houses of rich and poor alike in waterfront precincts. The affluent, who could afford to do so, moved farthest afield into new residential areas located on heights of land or oriented to some other amenity. Self-segregated suburbs formed, whose elevated social class and status was clearly implied. New "downtowns" also emerged, retail districts dominated by speciality shops catering to the well-to-do. There the new department store enhanced opportunity to shop, an activity that took on an empowering aura, especially for women of more affluent families who increasingly took responsibility for household buying. Meanwhile, women entered the workforce in larger and larger numbers, either as clerks in stores or as office workers.

Persistent change characterized American cities throughout the pre–Civil War era. Lightly and often crudely built of unseasoned wood and porous brick, city structures, by and large, were not meant to last. Or, they were not built to last at any one specific site. Typical of America in this era was the old building up on blocks and moving down a street toward a peripheral location, its vacated site made available for the higher rents and increased profits that come with greater development. The bulk of America's urban real estate was rented or leased, speculation in land and buildings accelerating as the nineteenth century progressed. The rectilinear grid of streets that typified most urban places was like a giant game board upon which business competitions were conducted and commercial successes and failures variously achieved. Salvaged buildings moved to the city edge, where land costs and property taxes were lower, to continue life as rental properties for the less affluent. At the periphery were farms awaiting the speculator's bid and also the squatters quarters.

These were the poorest of a city's residential sections except, perhaps, for the waterfront areas, where residual buildings not only sheltered the poor but also sustained prostitution, gaming, and other vices.

After the Civil War, the sorting and resorting of people and land uses accelerated dramatically. Extended horsecar, electric streetcar, and elevated and subway lines enabled the city to sprawl as never before. Affluent suburbs, physically removed from the city, evolved at station stops out along the railroad lines. In outline, cities assumed an overall star-shaped configuration, urban agglomerations orienting to the railroad corridors that radiated from city centers. Improved transportation fully inverted the city's zones of wealth and poverty. Increasingly, it was the affluent classes that took to the suburbs, now fully configured as pastoral retreats, antidotes to urban crowding and its associated noise, dirt, and threat of fire, disease, and social tension. Nearly every large American city suffered the extensive damage of periodic fires raging out of control over vast sections. Every city suffered cycles of epidemic disease, intensified by the lack of public sanitation. The new industrialism, which moved great wealth toward the top of the social hierarchy, produced frequent and sometimes violent complaint as the labor movement mobilized.

Increased prosperity and, over time, shorter work days, encouraged novel afterwork leisure activities. Traditionally, the theater and dining out in hotels, usually in late afternoon and early evening, had been substantially an elite prerogative. Moonlit summer nights encouraged strolling in public parks or along fashionable residential boulevards as well. The lesser classes also had their entertainments. As foreign immigration vastly increased, new kinds of popular leisure-time activity came to the fore. Germans, Poles, and Czechs brought the beer gardens and family-oriented taverns, drinking places very different from the largely male-only saloons long dominated by the Irish in America. The music hall was also introduced, and a wide array of organized sporting events were promoted. But, for the most part, city streets remained deserted after dark. Torchlight parades, lit bonfires on election night, and other after-dark diversions might be staged from time to time, but, all in all, the American city remained essentially a daytime place. The night loomed primarily as lost opportunity. The rise of affluence and increased leisure time, coupled with nighttime lighting, however, would change the urban night dramatically.

FIGURE I.1. "The Powers of Evil Are Fleeing before the Light of Civilization" is the caption given this woodcut, first published in the *Electrical Review* in 1885. With such illustrations, street lighting enthusiasts promoted the use of electricity as a means of enhancing safety in city streets. Street lights are depicted as lamp-carrying policemen standing vigilant against evil. From *Electrical Review and Western Electrician* 56 (May 21, 1910): 1053.

Night was traditionally the time when human activity slowed to re-energizing sleep, the hearth and shelter of home providing protection in this vulnerable time of poor visibility. Danger and evil were traditionally thought to lurk in the darkness of night. By cloaking criminals in darkness, night took on unsavory implication as something to be avoided in public streets. It was to combat this crime that systematic street lighting began. "Gaslight is the best nocturnal police," Ralph Waldo Emerson wrote.[6] Light disclosed. It offered surveillance. It gave city dwellers a sense of control over all that lurked in darkness, both real and imagined.

NEW YORK CITY

New York City certainly was not, and is not, the typical American city. It has always stood in a class by itself, sharing characteristics with the other world metropolises, like London and Paris. Nonetheless, New York City has always been the American style-leader, the trend-setter among cities. An innovation such as a new kind of lighting might not be introduced first in New York City, but, once adopted there, its acceptance was assured almost everywhere. Other American cities emulated New York's progressive cosmopolitanism. Linked to Europe through the nation's finest deep-water harbor, the city boomed after the completion of the Erie Canal, its western reach extending to the Great Lakes and beyond. Buffalo, Cleveland, Detroit, and Chicago developed as economic satellites closely knit to New York, first by water and then by railroad. Emulation of New York in these and other cities was often fostered by the direct investment of New York capital.

New York City itself quickly expanded northward along the length of Manhattan while also spilling into adjacent Kings, Queens, and Bronx Counties. Another of the nation's largest cities, Brooklyn, was absorbed, as great bridges were erected across the East River to tie a consolidated metropolis together. Lower Manhattan, however, remained the metropolitan hub. There the city was seen at its finest. There the principal institutions of commerce and government were located, increasingly housed in tall landmark buildings that spoke with force about the city's power as place. Located there also were the city's principal new entertainment zones. There the possibilities of nighttime lighting first entered American consciousness.

Broadway was, and remains, New York City's principal thoroughfare. After the Civil War, Broadway's southernmost stretch, from the Bowling Green north to Wall Street, was dominated by express agencies, the wharves of lower Manhattan being the traditional entryway into the city. From Wall Street to Ann Street insurance companies, banks, brokerage houses, and real estate agents dominated. Above St. Paul's Church and the Astor Hotel was City Hall and City Hall Park and, immediately to the east, Printing House Square, one side of which was Park Row, containing the offices of the city's principal newspapers. Beyond Pearl Street were the city's major dry goods stores, including Lord and Taylor and Brooks Brothers Clothing. Above Spring Street came the St. Nicholas Hotel, Tiffany's, the Metropolitan Hotel (behind which was Niblo's Garden, a pleasure ground), the Olympia Theater, the Southern Hotel, the New York Theater, and A. T. Stewart's Department Store at the corner of Tenth Street.[7] Broadway, therefore, was the first street in the city to be fully lit at night, first with oil lamps and then with gas and electricity.

Above Broadway's Grace Church in the late 1860s was Wallack's Theater at Thirteenth Street. Then came Union Square, formerly a fashionable residential area but rapidly giving way to commerce, with many townhouses being converted to business or demolished in favor of four-, five-, or six-story wholesale and retail buildings. "The windows of the stores are filled with the gayest and most showy goods. Jewels, silks, satins, laces, ribbons, household goods, silverware, toys, paintings; in short, rare, costly, and beautiful objects," wrote one visitor.[8] Above Fourteenth Street ran horse-drawn streetcars on tracks laid down in the center of the avenue. Below Fourteenth Street public transit was dominated by horse-drawn "omnibuses" or stages. "At night the many colored lamps of these vehicles add a striking and picturesque feature to the scene," wrote another observer.[9] Broadway was crowded through the early evening hours. A poem in *Harper's New Monthly Magazine* read, in part:

> *The lights are lit in dwellings and store;*
> *In countless numbers, score upon score.*
> *Of those that crowded the brilliant mart*

Are gone to their homes in the city's heart;
In the crush and tumult, hurry and press.[10]

Fifth Avenue crossed Broadway at Twenty-third Street at Madison Square. To the north along Fifth Avenue, a show street of mansions was rising to house the more pretentious families among the city's elite. Here was New York City's grandest hotel, the Fifth Avenue Hotel. North from Thirty-fourth Street along Broadway the area gave way to new suburbs of working-class poor. Through the 1870s and 1880s a mixed district of coal yards, stables, small shops, saloons, and tenements evolved with vacant land interspersed. Land values were relatively low, especially when compared to the axis of "high rents" evolving parallel to the east along Fifth Avenue. But the stage was set for a rapid change which would sweep over the whole of "Midtown" by the end of the 1890s.

The 1857 ban on steam-powered locomotives on the streets of lower Manhattan forced the Vanderbilt railroad interests to locate Grand Central Station on Forty-second Street, some four blocks east of Broadway. With this, Forty-second Street became the most important of the city's wider "crosstown" thoroughfares. With the gentry of Fifth Avenue resisting development of rapid transit there, the new subway lines were directed up both Broadway and Seventh Avenue, making the crossing of Broadway, Seventh Avenue, and Forty-Second Street the city's most important transit hub. This was Longacre Square, renamed Times Square in 1904 with the construction of the New York Times office tower.

The bright lights of New York City's entertainment industry led the way in making Broadway the nation's brightest nighttime street. In the 1890s, New York City theaters were clustered on Second Avenue, in the Bowery, on East Fourteenth Street, on 125th Street in Harlem, and along Broadway from Union Square to Forty-Second Street. The largest concentrations were around Madison and Herald Squares, the latter where Broadway crossed Thirty-fourth Street.[11] By 1900, the stretch of Broadway between Thirty-seventh and Forty-second Streets was known as the "Rialto," a concentration not only of theaters but also of offices for booking organizations (prime among them the Theatrical Syndicate, which came to control some 500 theaters across the United States), theatrical agents and producers, scenery, lighting, and costume suppliers, and printing houses serving the entertainment industry. Clustered along the cross-streets were boarding houses, restaurants, and saloons catering to performers. Here land-use specialization in New York City, and in American cities generally, was clearly illustrated. Here was an example of economic specialization, mutually reinforcing activity translated into mutually reinforcing land use.

ENTER THE AUTOMOBILE

At first, motor cars and motor trucks came unobtrusively to New York and other American cities. They seemed to fit right in, replacing horse-drawn vehicles. Internal combustion engines and battery-powered electric motors were an obvious improvement over animal power, even despite the high costs of purchase and maintenance involved. Cars and trucks moved considerably faster than a horse's pace, at least when not constrained by traffic, and they were thought to be infinitely cleaner than horses. Owning one of these new machines was also a matter of great prestige. Initially, only the wealthiest could afford automobiles, and cars were utilized more as sporting devices than as means of transport. By 1915, however, entrepreneurs like Henry Ford and General Motors' William C. Durant, using large-scale assembly-line production techniques, not only improved the automobile's mechanical dependability through standardization but also greatly reduced its cost, making automobiles affordable to the middling classes. By 1920, American cities were overrun with a large number of motor vehicles, used especially for commuting and business travel and for goods delivery.

Automobile use promised the ultimate in travel convenience, freedom of movement beyond the temporal and geographical restrictions of mass transit. Very quickly, cars and trucks congested city streets, especially in big city downtowns. They filled traffic lanes, particularly during rush hours, usurping curbside space for parking through the day. Between 1900 and 1920, motor vehicle registrations increased from fewer than eight thousand to more than 8 million, a figure which, in turn, more than tripled by 1940.[12] A new profession arose—that of the traffic engineer. Through the 1920s and 1930s, solutions to traffic congestion were sought primarily through street widening and regulation of curbside parking, strategies that converted public streets almost completely to traffic lanes. Pedestrian use of the street was vastly curtailed, as was the storing and displaying of goods at curbside and on sidewalks, which were much reduced in size given street widening. Street life was, in short, made less diverse.

With curbside parking increasingly banned to speed traffic flow, off-street parking lots appeared and buildings were demolished to make way for car storage. Demolition produced much fragmentation as traditional city blocks eroded and buildings were torn down ad hoc. In cities with well-established mass transit systems, especially those with elevated trains and subway lines like New York City, Philadelphia, Boston, and Chicago, fragmentation was somewhat delayed. But in most cities in the 1920s traditional downtowns unraveled quickly to accommodate off-street parking, especially in the booming new metropolises like Detroit and Los Angeles. By 1960, the residual downtowns of literally every American city were surrounded by asphalted parking lots and punctuated at their centers by parking decks or garages if not open lots. Upwards of two-thirds of the space in the typical big city downtown was devoted either to auto movement or to auto storage. Public space had become largely "machine space."[13]

The automobile changed commercial thoroughfares. Along commercial strips originally oriented to streetcars, curbside parking and, later, the parking lot came rapidly to the fore. Soon traffic congestion made streetcars inefficient, and they came to be viewed more as a hindrance to the movement of autos than as a viable transportation alternative. Streetcars had been replaced by buses in most cities by 1960. Curtailment if not complete elimination of curbside parking, again to speed traffic flow, reduced the viability of older strips, inviting parking lot fragmentation there as well. Retail and other functions were displaced to new shopping centers that were fully automobile convenient. Commercial strips developed with automobiles in mind were spread out affairs, dominated by parking lots adjacent to or in front of businesses. Automobile-oriented subdivisions were nearby. Densely packed houses on strict rectilinear street grids gave way to sprawling, low-density housing on curvilinear streets. No longer was a quick walk to a corner store or to mass transit possible. Now driving to shop and driving to work were made mandatory.[14]

When observers wrote of the automobile city after dark, it was often from the vantage of a moving car—the city playing out in a sequence of images obtained through a windshield. Sinclair Lewis so oriented his readers to Zenith, the fictional American metropolis in his novel *Babbitt*. George Babbitt, the novel's protagonist, describes a typical evening's drive downtown from his suburban "Dutch Colonial." The "demure lights" of little houses contrast with the "belching glare" of a distant foundry. Lights were on, as well, in the neighborhood stores, "where friends gossiped, well pleased, after the day's work."[15] Then comes

> an enormous graystone church with a rigid spire; dim light in the Parlors, and cheerful droning of choir-practice. The quivering green mercury-vapor light of a photo-engraver's loft. Then the storming lights of downtown; parked cars with ruby tail-lights; white arched entrances to movie theaters, like frosty mouths of winter caves; electric signs—serpents and little dancing men of fire; pink-shaded globes and scarlet jazz music in a cheap upstairs dance-hall; lights of Chinese restaurants, lanterns painted in cherry-blossoms and. . . . Small dirty lamps in small stinking lunch-rooms. The smart shopping-district, with rich and quiet light on crystal pendants and furs and suave surfaces of polished wood in velvet-hung reticent windows.[16]

John Dos Passos, in *Manhattan Transfer*, likewise takes readers along a suburban thoroughfare toward the city center. New York City was not immune, any more than any other place, from the automobile and its changes. A red light and ringing bells halts traffic for a train. "A block deep four ranks of cars wait at the grade crossing, fenders in taillights, mudguards scraping mudguards, motors purring hot, exhausts reeking." Then the light turns green. "Motors race, gears screech into first. The cars space out, flow in a long ribbon along the ghostly cement road, between blackwindowed blocks of concrete factories, between bright slabbed colors of signboards towards the glow over the city that stands up incredibly into the night sky like the glow of a great lit tent, like the yellow tall bulk of a tentshow." As for the road itself—"Cones of light cutting into cones of light along the hot humming roadside, headlights splashing trees, houses, billboards, telegraph poles with broad brushes of whitewash."[17] The roadside could be read as a text, one of lit objects and spaces made visible in the night. Whereas the city center blazed its established dominance, the suburban highway merely suggested its growing importance.

LAS VEGAS

Las Vegas is no more a typical American city than New York City. Yet, given the influence of automobility's now nationwide restructuring of urban geography, Las Vegas stands important indeed. Las Vegas has come to be the nation's ultimate highway-oriented urban center. Las Vegas also may have replaced New York City as the nation's leading city of the night, built as an entertainment center in the Nevada desert—where no large city really has a right to be. It was, of course, legalized gambling that energized the Las Vegas economy, its roadside motels adding casinos and nightclubs, which evolved into massive resort complexes, ultimately spreading from the edge of town southward along the Los Angeles highway at regular intervals.[18] The nearby development of Hoover Dam supplied cheap electrical power to light the giant sign spectaculars that boldly announced the casinos in the night. These signs became the hallmark of the city's nighttime functioning, both along the Strip, built anew, and downtown along a reconfigured Fremont Street.

Originated as a station stop on the railroad between Los Angeles and Salt Lake City, Las Vegas evolved eastward on a grid of streets aligned with the tracks. The essence of the early twentieth-century small town could still be seen at the center in the early 1960s. But drive out Las Vegas Boulevard toward the south and the new automobile-oriented city came fully to view. Here was a new prototype strip-city, auto-oriented and built fully around leisure-time entertainment. Here was a new city that specialized in life after dark—a city conceived and built from scratch as a nighttime place. Located two miles from downtown, at the city limit, was the Sahara Hotel, the first of an array of giant casinos—the Thunderbird, the Riviera, the Stardust, the Silver Slipper, the Desert Inn—all spaced at intervals of approximately a quarter mile. At the bend, where the road straightened due south, stood the Sands, and beyond came the Sans Souci, the Flamingo, and the Dunes. The very few cross-streets were named for their nearest casinos and led off toward golf courses and other amenities at some distance through largely vacant land. Each casino complex was announced by its giant sign out front, a "vulgar extravagance" in the words of architects Robert Venturi, Denise Scott Brown, and Steven Izenour. Their book, *Learning from Las Vegas,* sought to establish that

America's new automobile-oriented cities were not places of random chaos but were ordered functionally and visually. Regularity and order could be perceived from rapidly moving autos, especially at night.[19]

The giant signs provided much of the conceptual frame. The low casinos, back beyond driveways and parking lots, stood, in comparison, only as "modest necessities," even covered with their own giant signs.[20] Stretches between the casino complexes were filled with unobtrusive and unpretentious buildings. This new Las Vegas involved a "violent juxtaposition" of buildings of different scale and, at night, of different levels of illumination.[21] Dominating the view were the casinos, less visible in between were the mundane gas stations, fast food restaurants, cleaners shops, and drive-in banks that could be found everywhere in America. At casual glance, the whole did seem chaotic, which was how many architectural critics had described it. At night Las Vegas was a linear array of moving, pulsating, blinking light that made little sense to those accustomed only to traditional city space, conceived from the point of view of the relatively stationary pedestrian. From a moving automobile, however, order could be discovered in chaos.

Peter Beagle's book, *I See by My Outfit,* describes a trip by motorbike across the continent. At night two adventurers approach Las Vegas "herded by shadows toward a distant meadow of light." "Suddenly the highway has tightened on us," the narrator confides, "and the city swallows us with a roar of brightness. I wonder if this is how a candle looks to a moth come in out of the dark."[22] Together they head for the old downtown area and its main street, now fully refit with its own giant signs in competition with the Strip. "A block ahead of us, illuminated on the roof of the Golden Nugget . . . stands a mechanical cowboy, huge as a balloon in the Macy's Parade, bowlegged, baboon-faced, one hand hooked in his belt, the other going up and down slowly to tip his hat. Timed to the tip of the coffin-sized hat, his voice groans out, 'How—dy—pod—nuh.'"[23] Just as quickly, they leave the city for Los Angeles. What they see in Las Vegas is fully expected. "I have been to so many movies, seen so many night-club acts televised, memorized and chanted so much show-business mythology, that I could probably draw a map of the strip without mislaying too many colorful casinos. The magnificently tasteless facades float by, shivering like mirages of cuckoo clocks and wedding cakes, of gingerbread houses and

funeral homes and Howard Johnson's."[24] Finally, the highway grows dark
as the light of the city falls away behind. Gone is what another writer, Tom
Wolfe, called the "electric-sign gauntlet" where neon explodes in "sun-
bursts ten stories high."[25]

THE AMERICAN CITY HAD CHANGED, and nighttime lighting was
both a product of, and a contributor to, the metamorphosis. A modern
city was emerging, with new ways of organizing work in the production
of goods and services satisfying to an increasingly affluent American public. With increased affluence came increased leisure time, precipitating a
shift in popular preoccupation from work to play. In the early-twentieth-
century American city, nighttime entertainment zones sprung up where
theaters and other venues beckoned in nighttime visitors with bright light.
New York City's Broadway evolved, especially around Times Square, as
the "Great White Way." New ways of encouraging the consumption of
goods and services, coupled with the new automobility (itself a form of
heightened consumption), fostered different kinds of urban space, especially the strip, as epitomized eventually by Las Vegas.

The coupling of nighttime illumination with rapid movement by automobile and its promise of enhanced mobility was fundamental. This link,
however, served first to enhance and then to diminish night lighting in the
American imagination. The lighting of public space at night in cities had
not only had utilitarian benefit, encouraging to auto use, but also celebratory and informative applications, benefits substantially diminished
and even negated by eventual overemphasis on street lighting. Street lighting helped modernize cities and was, itself, taken as a sign of modernity.
Nothing symbolized modern progress more than brightly lit city streets.
But the bright glare of streetlights reduced the visual effectiveness of electric signs best displayed against dark backgrounds. The visual impact of
floodlit buildings was similarly diminished. In the process, much of the
excitement of after-dark hours was lost. Much of the romance and mystery of the urban night evaporated, street lighting, in its excess, no longer
capturing and holding attention as spectacle.

We take nighttime lighting so much for granted today that we easily
forget how recent in human history the privilege of sophisticated night life
arose—freedom to engage opportunities and enjoy pleasures extended

from the day into the night as well as those fully new to the night. Advocates of progressive city planning sang the virtues of bright night light from an early date. "The dark streets through which the pedestrian formerly made at night an uncertain way, with his individual lantern, now glow at midnight as at noon," wrote Charles Mulford Robinson in 1903, then the nation's leading proponent of "scientifically" planned cities. "Observe how much the modern city is indebted not merely for comfort but for dignity and beauty to recent discovery and invention," Robinson argued.[26]

Public illumination, like technology generally, seemed to have its own forward momentum. More was always better. In that assumption, the modern age seemed to offer unlimited promise. But even as early as 1914, a few observers had begun to express second thoughts, at least about overly bright nighttime lighting. Could there be such a thing as too much progress? At night, could there be too much light? Had streets become too bright, the light overly exaggerated? "To the resident of a city who never wanders from his own fireside or main street, the lure of the gay white way idea is irresistible," wrote one critic, F. Laurant Godinez, a spokesman for the window display profession. "But those who travel, and notice half they see," he continued, "are unanimously of the opinion that there is no better way of robbing a city of its individuality at night than by the introduction of these ornamental columns and balls." The lighting engineer, he argued, "can think only in terms of standardization." Night lighting, produced city to city, was "tedium and monotony."[27] Godinez was concerned that lit display windows in stores, among other nighttime lighting effects, would be overwhelmed by the brilliance of modern street lights. Street lighting revealed too much. Was this true? Could some dark shadow in nighttime city streets be a good thing after all?

The story of night lighting in cities is a tale of options developed and variously adopted or rejected. It begins with description of the nascent technology, especially the different types of lamp involved. It then embraces application of that technology, starting with street illumination, the essential form of night lighting—the kind of night lighting configured to the scale of the landscape. Like every story rooted in history, it offers perspective. What might be learned of America's past illuminating of the night that can be useful to configuring nighttime landscapes today and tomorrow? What are the lessons to be had from city lights?

PART I

EVOLVING LAMP TECHNOLOGIES

The Development of Night Lights

CHAPTER ONE

Oil and Gas Lighting

In the distant past, landscapes functioned little or not at all in the night, or not in ways thought normal and appropriate. Darkness obscured visual surveillance and provided a protective cloak for deviant behavior. Darkness seemed to bring out the worst in humankind, making the night something to be avoided beyond the security of home. As technologies evolved to light the night, nighttime landscapes emerged and took on new meanings. New opportunities were defined. A visible night invited celebration, including the celebration of light itself. The novelty of light attracted and entertained. However, as nighttime lighting lost its novelty, it became a mere facilitator, defining not excitement so much as the ordinary. Nighttime lighting became for many, if not most, Americans something easily taken for granted.

Human response to the night has progressed through three stages—avoidance, celebration, and elimination. For most of human history, avoiding the night was the most important strategy in coping with darkness. But, with the rise of artificial illumination, light could be cast across the landscape itself, creating in the process new kinds of public places. The new light offered celebration if only to demonstrate the prowess of modernity in overcoming nature's night. Light was made tantalizing and inherently entertaining. So successful was new lighting that Americans even began to believe that night could be eliminated—that the night could be turned into day in a visual and thus functional sense. Light could be used to standardize and homogenize places across the diurnal cycle. The development of oil and gas lamps represented initial steps in that direction.

Through the seventeenth century, the only lights on most city streets in both Europe and North America were the watchmen's lanterns or the bonfires they maintained about guardhouses. In the eighteenth century, however, authorities in many cities allowed ordinary citizens nighttime rights of traverse if they were carrying lanterns. Then came stationary lanterns positioned at intervals along principal streets. New York City required that lanterns with candles be displayed before every seventh house, expenses for each light to be shared by the households immediately served. The aldermen of each city ward were to see that lamps were lit "in the Dark Time of the Moon."[1] Candles were made of a variety of animal and vegetable fats, especially tallow and suet—byproducts of slaughtered sheep and oxen. The best candles, however, were made from spermaceti obtained from the heads of sperm whales. This tasteless, colorless, sponge-like substance, when pressed, proved unrivaled in delivering steady, clean, long-lasting flames.[2] A premium product affordable to only the wealthiest households and the most profitable businesses and used by city governments for their most important public lighting, spermaceti candles remained the standard for candlelight for over a century. A spermaceti candle was chosen as the standard for measuring luminance, or candlepower.[3]

Candles were improved with the introduction of plaited wicks and petroleum-based oils to extend candle life. Candles continued to serve domestic indoor purposes better than public outdoor needs. Candles fared poorly in high winds, and they cast relatively little light. Candled street lamps barely outlined the form of streets; people moved from light to light as from one feeble beacon to another, a sense of darkness still prevailing. Henry Thoreau wrote with a sense of irony: "Men are generally still a little afraid of the dark, though the witches are all hung, and Christianity and candles have been introduced."[4] Lamps with candles held promise but did not meet the needs of nighttime lighting.

It will be helpful at the start of this exploration of night lighting's landscape implications to think of lighting as more than mere illumination. Lighting technology, in general, moved toward the brightening of the American night. Let us consider, if only briefly, the kinds of lighting technologies involved before emphasizing their applications. Each lamp technology illuminated the night differently: each burned at different levels of brightness or luminance and with different hues or levels of color saturation, among other variables. Costs of manufacture, operation, and maintenance varied from one kind of lamp to another, determining the market feasibility of each. To appreciate why various lamp technologies were adopted and to appreciate what visual, and thus functional, effects they wrought, a basic understanding of lighting's technological base is required. I start at the beginning, with oil and gas lamps.

OIL LAMPS

Until the age of gas illumination, the most important street lighting improvements involved oil lamps. Lamps burn many of the same oils from which candles were made, but they did so much more efficiently. Lamps draw oil through a wick from a reservoir. When lit, the flame converts the liquid into gas for combustion, just as a candle flame converts tallow first into a liquid and then into a gas. Lamps produce more gas faster, heating the wick to higher temperatures and, in the process, producing more intense light.[5] The quest for ever brighter light would remain a prime objective of lighting innovation over the years. Combustion in oil lamps is imperfect; free carbon escapes as smoke, blackening lantern glass and dimming brightness. Mirrorlike lamp refractors and glass lamp surrounds required constant cleaning. Also demanding constant attention were the refilling of reservoirs and the trimming of wicks, making oil lamps very labor intensive. The drive to lower costs in the servicing of lamps would prove another enduring stimulus to innovation. After servicing the lamps of his route during the day, the lamplighter returned each evening at dusk to separately light each fixture. Friction, or locofoco, matches—pine wood shavings with melted sulphur tips—were perfected in the 1830s as safety matches through the addition of red phosphorous. Carrying matches or a torch, the lamplighter marched along lamp to lamp in predictable progression.

Benjamin Franklin contributed substantially to oil lamp innovation in the 1750s. He surmised correctly that two wicks and two flames burning close together would provide more updraft, increasing brilliancy in lamps. He introduced a loosely braided wick, which made for more efficient capillary action, and added air vents at the bottom and a funnel at the top to further increase the draw of air, improving combustion and reducing

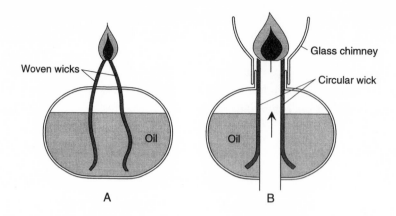

FIGURE 1.1. Oil Lamp Improvement in the Eighteenth-Century. Benjamin Franklin's lamp involved two wicks to siphon oil from a reservoir (A). Aime Argand's lamp used a hollow wick to enhance the draft of air, aiding combustion (B). Illustrated here are the basic principles upon which more sophisticated gas fixtures evolved. Modified from Harold F. Williamson and Arnold R. Daum, *The American Petroleum Industry: The Age of Illumination, 1859–1899* (Evanston, Ill.: Northwestern University Press, 1959), 32. 1959, © By Northwestern University Press.

smoke even more. Franklin's lantern used flat panes of glass, which could be easily cleaned and replaced and were less expensive than the blown glass globes imported from London. Swiss physician and chemist Aime Argand further enhanced lamp technology in the 1780s. Argand fit a glass chimney over the lamp flame and substituted a circular or tubular wick, which vastly improved combustion as air was drawn up through the center of the wick as well as around its outer edges.[6] Later Argand discovered that constricting the lamp chimney near the height of the flame increased air turbulence, further amplifying lamp brightness. In 1800, the Carcel lamp was introduced, with a double piston operated by clockwork to force oil through a tube toward the flame.[7]

Lamp oils derived from a variety of sources. The oil of the Greenland or arctic right whale became commonly available once the whaling industry had organized to tap the resources of the North Atlantic. But, as with candle manufacturing, the oil of the sperm whale was preferred, extending whaling, as a result, far into the South Atlantic, the South Pacific, and beyond. Spermaceti oil was used in New York City's street lamps as early as 1792.[8] The improved oil lamps were brighter than candles and created stronger light at intervals along darkened streets for pedestrians. Travelers commented on this new feature of the city streets in diaries and journals. At Alexandria, Virginia, one anonymous diarist penned: "[The streets] are lighted every dark night. A man, or perhaps more, goes round at dusk with a ladder in their hand, by which they ascend the post, and set fire to the lamps. These lamps are at every corner where the streets cross. The lamp is placed in a large glass lantern, such as taverns use and this is tenaciously fixed on the top of a high post out of reach, so that disorderly persons may not bring in their power to extinguish them."[9]

Lantern smashing, which reached a pinnacle in the French Revolution, was an act against authority. The destruction of street lights erected a wall of darkness, reducing the power of authority.[10] That lighting was a device of police control in Alexandria was little disputed by the diarist. Instead of bells, the writer continued, the night watch was announced by trumpet calls through to ten o'clock. Persons who could not account for themselves were placed under arrest in a guardhouse until morning, at which time they were brought before the magistrate of the city and fined.[11] It bears repeating that as light enabled surveillance, it offered the appearance of control. This assumed relationship between public lighting and the discouragement of crime would stand as accepted truth through the nineteenth and into the twentieth century.

Boston's street lighting began in 1773. A newspaper editorial the previous year had championed a system of public lamps so that those who ventured forth at night on calls of business or friendship might be protected from "insult, abuse, and robbery." Lamps, it was argued, would deter "scenes of lewdness and debauchery which are so frequently committed with impunity at present."[12] A citizens' committee, chaired by John Hancock, oversaw the placing of more than three hundred lamps brought from England. London's experience informed the Boston venture, and the rules for lighting London's streets were adopted. Lamps with globes ten inches in diameter were placed on posts ten feet high and some four feet out from building facades. These were distanced fifty feet apart along city streets.[13] Even still, illumination remained more suggestive than real, pedestrians moving from one pool of light to another through intervening shadow.

New York City streets were lit by more than 1,600 oil lamps in 1809. With 1,100 lamps, Philadelphia, more so than Boston, followed in rivaling the New York system.[14] It is important to remember that oil lamps persisted even after the introduction of gas technology. Indeed, the lighting of every era involved holdovers of old lighting technologies in combination with current innovations. In 1834, seven years after the first gas lamps had been installed on New York City's Broadway, the city counted only 384 gas street lights, while there remained more than two thousand oil fixtures. Indeed, the number of gas street lamps did not exceed that of oil lamps in New York City until 1851.[15] In part, this oil lamp persistence reflected the development of better and cheaper illuminating oils.

In the 1840s, camphene (also called spirit gas), a mixture of turpentine and alcohol, came onto the market. Much cheaper than previous oils, it was adopted widely among the poorer classes and substantially democratized domestic lighting. Merchants also widely used it in lighting retail stores. Camphene's light weight and low boiling point was advantageous. The fluid moved more readily through lamp wicks, particularly in cold weather, and it produced a strong white flame, which greatly enhanced lamp brilliance.[16] Camphene was the first synthetic lamp oil used in the United States, and its distillation directly led to the development of fuels derived from bituminous coal and petroleum. "Mineral oils," similar to present day kerosene (which itself is still sometimes improperly called "coal oil"), were distilled from coal tar as byproducts of artificial gas production. It was, however, not until the discovery of vast quantities of petroleum, beginning with a find at Titusville, Pennsylvania, in 1859, that lamp lighting was revolutionized with actual kerosene.

Crude petroleum separates into a number of products through fractional distillation. From the light to the heavy, in terms of densities or weights per gallon, came gasoline, kerosene, and lubricating oils, with paraffin and petroleum jelly being near solids. Kerosene is less explosive than camphene, but it is dirtier and burns with an odor. Yet, kerosene was also much cheaper to produce and fell dramatically in price when supplies glutted markets with the rise of the nation's railroad network. Kerosene lamps produced 10 to 20 candlepower, one gallon of kerosene yielding upwards of eight thousand lumens.[17]

The use of reservoir-filled lamps in street lighting culminated with the

FIGURE 1.2. Advertisements for Gasoline and Gasoline Street Lamps. Oil lamp technology ultimately led to gasoline fixtures, still seen in American cities as late as the 1920s, especially on side streets and back alleys, where brightness was not as necessary. Cheap gasoline for automobiles translated into cheap fuel for lamps as well. From *Gas Age* 31 (May 1, 1913): xxvi.

adoption of gasoline as a fuel (see fig. 1.2). Gasoline, its price driven down by the supply produced for widespread use in automobiles, was employed extensively after 1900 in old lamps converted from kerosene. Gasoline lamps were used in the twentieth century to light alleys and lesser streets at urban peripheries where electrical lighting was not yet economical. In 1926, Chicago employed 5,500 gasoline street lamps (compared to 58,500 electric lamps); Philadelphia had over 11,300 (compared to 27,400 electric fixtures).[18]

GAS LAMPS

Many historians assign revolutionary implications to gas lighting. The creation of gas factories with pipeline systems for gas distribution is seen as ushering in the era of giant public utilities. Later, the generation and supply of electrical power through similar, centrally focused networks only echoed the gas industry's earlier accomplishments. Frederick Collins, a historian of New York City, credited early gas technology with changing "darkness into light," thus making possible America's largest metropolis. Collins considered gas lighting more important than the Erie Canal or Fulton's steamboat.[19] It was oil lamps, however, that had led the way in public illumination. Gas lighting emerged only after it had become, through scale of operation, a more efficient and less costly method for delivering combustible vapor to lamp fixtures. The rise of gas lighting reflected the availability of coal brought by steamship and later by canal and railroad. It was with cheap coal, imported first from Newcastle in England and later from coalfields in Virginia and Pennsylvania, that gas, and thus gas illumination, spread in America's eastern cities.

Gas had been a discarded byproduct of coke production. Produced in the iron retorts in which coal was superheated into coke, gas had merely been released to the air. The capturing of gas exhaust in containers, and its use in commercial lighting, first occurred in 1802 at the Watt and Boulton cotton mill near Birmingham, England.[20] Gas was directed down metal tubes and lit at connecting lamp fixtures. William Murdock worked at filling bags made of leather and then of polished silk, in attempts to create lights as conveniently portable as oil lamps. He settled instead for systems of permanently fixed gas retorts, storage tanks, or gasometers, and pipes and valves that created a network of lamps at the scale of the factory. The first lighting of a public street by gas occurred in London's Pall Mall five years later.

Americans had their first glimpse of the new gas technology in 1816, when Charles Kruger used methane—carbonated hydrogen gas produced from wood tar—to light a portion of Peale's Museum located in the Old State House in Philadelphia.[21] Within weeks the technology was demonstrated in Baltimore at the Peale Museum there, and within a year an investor syndicate organized the Baltimore Gas Light Company. This was the first American system for lighting public streets with gas. Philadelphia would not adopt gas for public lighting until the 1830s, and even then it was only after bitter public debate, opponents fearing explosions and fires (and higher insurance rates, accordingly), as well as contamination of water mains. Most telling, perhaps, Philadelphians objected to the constant digging up of city streets.[22]

Experiments with gas lighting were conducted in 1817 in New York City, but, as in Philadelphia, they failed to prompt immediate commercial application. The oil-based lighting systems of New York and Philadelphia were large and sophisticated enough to preclude quick replacement by a still unproven alternative. In 1824, after a gas street lamp had been placed in New York's Franklin Square as temporary demonstration, a journalist reported that "although only a small burner has been placed in the lamp any person may comfortably read a newspaper by it."[23] That same year, Samuel Leggett, a director of New York City's first gas company, installed at his home the first permanent domestic gas installation. Another journalist commented: "each room . . . was literally illuminated by the most vivid and pleasant light we ever beheld." "The effulgence of these lights have a wonderfully fine effect upon the human countenance," he continued, "by affording it a youthful brilliancy and adding splendor to every surrounding object."[24] By the end of 1825, Leggett's company had some six miles of gas main radiating out from a gas house, one of the city's largest structures, on Hester Street. The gas system by then served nearly six hundred houses and shops with more than three thousand gas lamps.[25]

Permanent gaslight installations were placed on New York City's Broadway in 1827. Broadway, the city's most important thoroughfare, follows the highest ground between the East River and the Hudson. It runs in a straight line from the Battery to Tenth Street, where it curves northwestward as a diagonal cutting across Manhattan's street grid toward the city's outskirts (then only a few blocks away). Eighty feet wide, the street was lined with some of the city's most elegant houses and shaded in stretches by tall poplar trees. Although primarily a residential street, here also were located many of the city's most exclusive shops. Below city hall and west of Broadway was New York City's most fashionable residential district. East of Broadway were the immigrant neighborhoods oriented toward the East River docks. Broadway was not only a major axis for movement, it was also a major social dividing line.

In 1827, 120 iron lampposts were erected along Broadway from the Battery to Grand Street and fitted with gas lights. They joined 3,100 oil lamps already in service on city streets.[26] In succeeding years, gas lights were erected on Chatham and Pearl Streets near city hall and then before city hall itself. Frances Trollope, visiting New York City in 1829, observed: "At night the shops which are open till very late are brilliantly illuminated with gas, and all the population seems as much alive as in London or Paris."[27] By 1835, 384 of the city's 5,660 street lamps were gas.[28] Street lighting, as indeed was the case with gas distribution generally, involved city franchises. In 1856, the New York Gas Light Company serviced the city south of Grand Avenue and supplied some three thousand street lamps; the Manhattan Gaslight Company serviced the city north of Grand Avenue and provided some 7,300 street lamps.[29] In 1893, New York City streets were lit by 26,500 gas lights and only 1,500 electric lights.[30] Numerous companies continued to manufacture traditional gas lamp fixtures up through the early twentieth century.

Making gas from coal followed a process of destructive distillation. When coal was superheated in a retort, it fused as coke, releasing hydrocarbons from which color dyes and other products can be made. The coke itself was sometimes used in the production of "water gas." Steam was forced through the red-hot coke, producing hydrogen and carbon monoxide, which were then enriched by mixing oil into the steam.[31] The gas works of the Manhattan Gaslight Company was located on filled land at the foot of Fourteenth Street along the East River. It comprised three retort houses, each about 240 feet long; six large gasometers, each about 97 feet in diameter and capable of holding roughly 368,000 cubic feet of gas; buildings housing scrubbers and condensers; an engine house; a machine shop and blacksmith shop; an office building; and a large coal yard.[32]

There were other gas illuminants developed. A byproduct of iron and steel manufacture, calcium carbide, produces acetylene gas when water

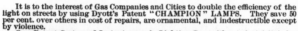
FIGURE 1.3. Advertisements for Gas Lamp Fixtures. Among the lamp brands advertised in 1888 were Bartlett, Miner, and Standard, each manufacturer offering an array of styles. From *American Gas Light Journal* 49 (Nov. 16, 1888): 363.

is added to it. Here was a gas that could be made in small quantities literally anywhere. Due to its convenience, acetylene lighting came into use
on ships, railroad trains, and, later, in automobile headlights. Acetylene
produced an exceptionally bright white light, much like the limelight made
by a similar process as early as the 1820s. Limelight was produced from
lime, later replaced by oxide of zirconium (or zirconia), heated in the flame
of a Bunsen burner to produce light upwards of 400 candlepower. Limelight was used in projection lanterns and, more memorably, in theater

spotlights—hence the more colloquial meaning of *limelight*. Limelight and acetylene were soon used to light searchlights and floodlights in outdoor illumination.

Traditional notions of free enterprise—whereby free competition in an open marketplace prevailed—could not be fully reconciled with large gas lighting systems. Whereas in some cities competing gas companies laid their mains side by side in the streets, demand usually could not support such redundancy. Sometimes public rights-of-way simply could not physically accommodate the mains of all competitors. Therefore gas companies, like the mass transit companies, were given territorial franchises under governmental supervision. In theory, government ensured that individual companies coordinated operations and otherwise served the public good. Gas quickly became big business. The eleven gas companies of the five largest American cities had invested by 1859 some $11.5 million in operations, producing an annual output valued at $3 million.[33] As an industry, only the garment manufacturers in those cities exceeded that level of capitalization.

Gas technology came later to western cities. Gas lighting in Chicago coincided with the 1848 opening of the Illinois and Michigan Canal, which made cheap coal readily available. What would become the People's Gas Light and Coke Company was formed by a group of city boosters in 1850. A municipal franchise to lay pipes beneath Chicago's streets was secured, with city government subsidizing the effort through the purchase of street lamps and contracts for street lighting services. Gas lighting in Chicago, as in eastern cities, was made available primarily in the central business district and in more affluent residential areas. Initially, only upper-income families could afford to light private dwellings with gas, just as only they could afford the special tax assessments that city governments levied to light residential streets. Monopolistic practices retarded the extension of services geographically, as gas companies generally refused to lower rates in line with declining costs of coal. At least 70 percent of the price charged to gas customers in Chicago in the early 1880s represented profit for stockholders.[34]

Gas light was clearly better than that provided by oil lamps. Up to ten times brighter than oil light, it appeared to some so intense that shades of frosted glass were frequently employed to produce a more amorphous or diffused light. Of course, opinions differed. The anonymous traveler to Alexandria also visited Philadelphia and wrote there: "It is at night that the wealth and splendor of Philadelphia appears to the best advantage; the windows being lighted with numerous lamps and gas-lights, also with the lamps in the streets, and the luster of the pleasing wares in the windows, present a scene of astonishing beauty."[35] But Englishman James Buckingham, traveling in the United States on a lecture tour, found New York City's gas lights inadequate. "The lamps are so far apart and so scantily supplied with light that it is impossible to distinguish names or numbers on the doors from carriages or even on foot without ascending the steps."[36]

Cities function with difficulty at night without street lights. When an 1848 fire destroyed the Hester Street gas works, many of New York City's major streets were left in the dark, people stumbling into one another and into things. Yet most sections of nineteenth-century New York City were never lit by gas. Along the East River, for example, wharves and piers built of wood (rather than of stone, as in Europe) were too unstable for gas mains. Oil lamps also continued to provide what meager illumination there was in the poverty-stricken immigrant districts. At night these areas stood in stark contrast to the best gas-lit portions of the city, especially Broadway's theaters.

Theaters were among the first private facilities to be lit by gas, and, with the later arrival of electricity, they were the first to be lit by electric light as well. Other important early users of gas lighting included pleasure gardens, taverns, hotels, museums, and department stores. All had some sort of nighttime entertainment implication catering to large numbers of people. In 1816, the New Theater in Philadelphia became the first commercial building in the United States to have its interior lit totally by gas.[37] Other large theaters with substantial lighting needs included the Park in New York City, with 2,500 seats (organized in three tiers of boxes and a large gallery), and the even larger Bowery Theater, also in New York, with 3,500 seats. Both were opened in the 1820s. At first, only the theater stages were lit with the new lamps, but the boxes occupied by society's elites followed. As one newspaper editorialized in 1824: "What is the use of spending two hours at the toilet if no one can perceive the improvement. . . . Unless the managers correct this evil we would suggest that every lady like the ancient vestals of the East bring her lamp."[38] Candles created problems in lighting theater stages. Few plays went uninterrupted by the "snuff boys," whose responsibilities included replacing candles burnt or blown out. Actors spoke their lines while stepping into and out of the focused light, often exaggerating their gestures in order to be seen. Gas lights enabled more natural acting, and, as the entire stage could be used with the new lighting, more elaborate scenic backdrops came into use.

The owners of the Castle Garden at the Battery in New York City inaugurated gas lighting in 1825. A newspaper advertisement read: "There will be from four to five hundred jets of light attended with a first rate Band of Music."[39] A year later gas lamps were introduced at the Vauxhall Garden on New York City's Bowery Road. The first use of gas as an advertising medium occurred at Niblo's Garden at Broadway and Prince Street in 1828. On the opening of a theater on the grounds, the proprietors encased gas lights in "glass cups" to brightly spell out the word NIBLO in red, white, and blue letters.[40] P. T. Barnum adopted gas light both inside and outside his American Museum on Broadway, while limelight was projected from the roof of his building. A band played every evening from a balcony, light and sound blatantly orchestrated to attract patronage.[41]

Through the Civil War and beyond, gas lighting proved essential to a variety of distinctive settings or places in American cities—places that sold mood as well as product or service. Hotels, such as the giant Fifth Avenue Hotel in New York City, used sedate gas lighting as an inducement to pleasure. The large, brilliantly lit parlor on the second floor was celebrated as "a field for fashionable flirtation after dinner."[42] Such rooms gave New York City a decidedly cosmopolitan air. So too did the city's larger retailers who adopted gas lighting, especially the new department stores.[43] If hotels were so-called palaces for the traveler, then department stores were palaces for the shopper, especially affluent women shoppers. Gas was an important tool of large-scale merchandising, a lighting technology capable of brightening up the interiors of vast, new buildings. An aura of security and splendor was promoted.

The facade of A. T. Stewart's store was kept free of signs, unlike Barnum's American Museum. Gas light enhanced the architectural grace of the building and illuminated rows of large sidewalk windows which were fully suggestive of luxury. Gas lighting's relative brilliance could excite and stimulate better than the light of oil lamps, and, in so doing, it could appeal across the full range of society. Nonetheless, gas lighting could also be used pretentiously to enhance a sense of elegance to which the more affluent alone were attracted. Gas suggested modernity, just as electricity would later. Embracing it was a sign of being up-to-date and fashionable.

As he had been during the era of oil lamps, the lamp lighter remained an important presence on America's streets in the gas age. Not until the perfection of electrically controlled pilot ignition systems, after 1900, was the lamp lighter made obsolete. Author Robert Louis Stevenson was an advocate of gas light. Although it shown with some brilliancy compared to oil illumination, it did not impose the oppressive glare that electric arc lamps brought to public places. Thus Stevenson made the lamplighter a romantic icon of a rapidly vanishing past, a past giving way to excesses in electric lighting. The more leisurely pace of life symbolized by the gas flame seemed to be disappearing.

> My tea is nearly ready
> and the sun has left the sky;
> It's time to take the window
> to see Leerie goin by;
> For every night at tea-time
> and before you take your seat,
> With lantern and with ladder
> he comes posting up the street.[44]

In New York City, lamplighters were usually appointed and paid by the gas companies, under direction of the city's Superintendent of Lamps and Gas. Each man was given one hour to light the lamps of his district. Tools of the trade included a cleaning ladder, a lighting ladder, a pair of tongs, a pair of pliers, a brush, and, of course, matches. Matches were replaced by long electrically charged rods toward the end of the nineteenth century.

Gas lighting did not simply disappear with the advent of electricity. Many cities initially retained gas lamps even on streets where electric light had been installed. In Philadelphia in the 1880s, electric lighting was viewed as an experiment, a technology not yet proven. "The city still keeps gas burning—that is, feebly flickering—on streets where the electric

FIGURE 1.4. A Lamp Lighter. After electric street lights came to the fore and electrical starters were widely employed on gas fixtures, the lamp lighter was cast as a romantic figure. Traditional gas lighting was dim, especially compared to the electrical lighting that followed. Nostalgia for gas lighting retrospectively imbued older cities with an aura of romance. From *Electrical Review* 5 (Jan. 24, 1885): 3.

light is used, for fear of accident to the electric conductors or machinery," noted journalist Arthur Bright. "But we are told that this course was followed half a century ago in regard to oil lamps when gas was first introduced."[45] Still, advanced technology continued to extend gas use, especially the Welsbach gas mantle, developed in Austria (see fig. 1.5). A more brilliant flame was produced, requiring only two-thirds the gas formerly needed. Efficiencies were raised from between 4 and 6 candlepower per cubic foot of gas per hour to 60 and 70 candlepower. Containing the flame within a sealed container, as with the incandescent electric bulb, kept the cost of gas illumination competitive with electric light. Baltimore still operated more than 10,000 of these fixtures as late as 1954.[46] Gas remained a form of efficient incandescence, but it used a gas flame rather than a filament set glowing by electrical current.

Bolstered by seemingly unlimited supplies of natural gas, especially after the opening of gas fields in Pennsylvania, West Viriginia, Oklahoma, and Texas, gas companies turned to markets other than lighting. Gas utilities redefined themselves as suppliers of fuel for cooking and heating. Beginning in the 1950s, gas lighting enjoyed a very modest revival and was applied to illuminating entertainment districts such as San Antonio's River Walk, improved for the Hemisfair of 1968. Gas lighting also appeared in new upscale suburban subdivisions and in central city neighborhoods revitalized through gentrification. Gas fixtures, by night or day, suggested times past. Flickering gas light created a mood and implied an elevated status rooted in the nostalgia of a romanticized past.

IT IS DIFFICULT NOT TO CONCEPTUALIZE oil and gas lighting as merely preliminary to the use of electricity. Certainly, despite the lingering use of gas lighting even today, electric lighting has proven overwhelmingly to be America's illumination of choice. Nonetheless, in looking back, one must be careful not to diminish the accomplishments achieved with earlier technologies. With oil lamps, public lighting became an indispensable aspect of modern urban life. With gas lamps, the potential for night lighting in public places was largely fulfilled. Electricity would supplant gas only after it had proven not just brighter but, importantly, more economical. Illumination by both oil and gas had downsides, however. Neither oil nor gas were fully suitable for indoor use, as they

FIGURE 1.5. Advertisement for Welsbach Gas Lamps. The gas industry sought to keep gas street lighting competitive with electrical illumination by developing incandescent lamps with sealed mantles. Ultimately, however, the gas utilities would focus on the domestic market, promoting gas for cooking and heating. From *Gas Age* 31 (Jan. 1, 1913): xxxi.

were dirty, unhealthy technologies. Gas lights, for example, consumed enormous amounts of oxygen, requiring systems of ventilation that few theaters, hotels, and other large places of indoor assembly could provide. Attending the theater gave people headaches and even nausea owing to the accumulation of ammonia, sulphur, and carbon dioxide in the air.

Yet urban night life came of age in the era of oil and gas lighting. Oil and gas lamps allowed, and encouraged, the pursuit of leisure activities after dark in public places. For men especially, the night was enhanced for pleasurable adventure in saloons, music halls, brothels, and gambling halls. But retailers and other more "legitimate" businesses also embraced the night. They did so in conjunction with the polite entertainments of the theaters, hotel dining rooms, restaurants, and pleasure gardens. The new department stores especially shined in the night as an invitation to the nation's increasing materialism. Stores stayed open into the evening hours sometimes several days a week, and even when stores were closed their lit display windows encouraged "window shopping," a new form of leisure activity.

Although Europe, notably the capital cities like Paris, London, and Vienna, led the way in illuminating public places with oil and gas lamps, America kept pace. In New York City, Boston, and other cities, principal commercial streets were lit to aid the circulation of people and goods in city streets after dark. Americans were perhaps less involved in the lighting of public squares as ritual or ceremonial places, if only because American cities contained relatively few of these. It was the rectilinear grid of streets that dominated American cities after 1800, and Philadelphia was planned on such a grid from its very beginning. In America it was the street that dominated the lighting engineer's work.

Nonetheless, streetlights did not overwhelm other kinds of lighting in downtown commercial districts. Contributing substantially to the illuminated night and often dominating what observers saw and remembered were the lights of display windows and, for their novelty, the lit signs over doorways. What is most important in terms of the changing landscape, however, is the fact that light took meaning in contrast to backgrounds of dark shadow, even if the flickering lights of oil and gas barely relieved the darkness. The power of light stood in stark contrast to surrounding gloom. Light facilitated city functioning after dark, but it could still excite the mind as being more than merely utilitarian. Prevailing darkness continued to engage the imagination, sometimes in romantic speculation and sometimes in fearful apprehension.

CHAPTER TWO

Electric Lighting

Electricity, long known but little understood, seemed in the early nineteenth century ready for practical application as an illuminant. Electricity was first identified as a force in nature by the Greeks, who observed that amber rubbed with a cloth would first attract and then repel small metal objects. William Gilbert elaborated this experiment two thousand years later. While serving as physician to Queen Elizabeth I, Gilbert outlined the fundamentals of magnetism, describing the earth as a huge magnet and naming its mysterious attractive force *electrica* after the Greek word for amber.[1] Others differentiated electricity from magnetism, such as Otto Von Guericke, who created an early generator capable of producing sparks from friction, and Francis Hauksbee, who produced a glow in an evacuated hollow glass globe. Physicist Alessandro Volta perfected battery cells for storing electrical charge, pointing the way for chemist Humphrey Davy in 1809 to connect two pieces of charcoal to a primitive battery and create a brilliant arc light.[2]

Experimentation with electricity and electrical lighting accelerated in the early nineteenth century. In 1831, Michael Faraday in Britain and Joseph Henry in the United States both independently discovered the principle of electromagnetic induction, leading to the development of the dynamo as an electrical power source. Early experiments with arc lamps failed to produce long-lived light. Experiments with crude incandescent lamps using a variety of filaments borrowed from arc lighting—such as charcoal, carbon, and graphite—also proved impractical.[3] In the 1870s, however, arc lights for lighthouses and other uses were perfected in Europe after a number of technical obstacles had been surmounted, such as finding a sustained supply of improved carbon, developing mechanical regulators (to maintain the proper alignment of the burning carbon rods), and applying superior dynamos.

The arc lamp involved a very high voltage stream of electricity—actually a flow of opposing negatively and positively charged ions—between two electrodes. Light was not produced by the electric arc itself, as was at first surmised, but by the heating of the carbon electrodes by the electric charge. Simultaneously, some light resulted from combustion when carbon particles were released and smouldered in the air between the electrodes. Unlike the incandescent lamps subsequently developed, which are contained in airtight containers, open arc lights actually burned in the open air, much like a candle flame consuming its wick. The greater part of the light was generated in a white-hot crater formed in the anode, or positive post. The interior of the arc was violet in color, while its extreme surround was a greenish yellow. The earliest arc lamps generated up to 800 candlepower. Deposits that built up on the negative post interfered with arcing and produced the hissing sound long associated with electric arc technology.

In Europe, the standard arc lamp, called the "electric candle," was developed by Russian inventor Paul Jablochkoff. Jablochkoff placed two rods of carbon side by side, separated by gypsum insulation. Lit at the top, the candle burned down and required mechanical regulation. Usually four of these devices were placed in glass globes some twenty inches in diameter. As each candle burned out, the electricity was switched automatically from candle to candle until all were consumed over a period of several hours. Like gas fixtures, arc lamps required daily servicing and were, accordingly, very labor intensive. Nonetheless, they were an intensely brilliant source of light.

EARLY APPLICATIONS

Jablochkoff lamps were first used in England in 1878—in a London viaduct, at London's Gaiety Theater, and at a Sheffield sports stadium, where more than 30,000 people gathered to witness the first soccer match played at night. Light was thrown from four lamps elevated some thirty feet above the playing field.[4] The electric candle became most associated with Paris, however, in the public mind. During the Universal Exhibition of 1878, both the Place and Avenue de l'Opera were lit by arc lamps set approximately every 150 feet, each electric fixture replacing up to six gas fixtures.[5] An observer wrote of the splendor of the place: "The effect is magnificent and at this moment there exists nothing in this city of splendid effects to compare with this magical scene. The vista is about two-thirds of a mile, and the effort incomparatively finer than any show of artificial illumination ever before seen."[6] Thus Paris reinforced its appellation as the "City of Light" through the new electric arc technology.

The first public display of arc lighting in the United States took place indoors at the Philadelphia Centennial Exposition in 1876. At the time, the still-immature technology was vastly overshadowed by George Corliss's thirty-foot, 1,400-horsepower steam engine, which would come to symbolize for Americans the apogee of the steam age. The first American outdoor demonstration of arc lighting occurred in Cleveland in 1879. Charles F. Brush experimented with arc lights, producing a sophisticated system of lights wired in series, but with automatic shunts on each lamp. So if one lamp failed, none of the others were affected, a feature that previous systems had lacked. Brush also improved lamp regulators, adopted copper-plated carbons, and, as with the Jablochkoff fixtures, adopted multiple candles for all-night illumination. He also created an improved dynamo and an improved storage battery. Brush contracted with the Telegraph Supply Company of Cleveland, giving that firm exclusive rights to manufacture his equipment for a royalty fee. Within a few years the firm changed its name to the Brush Arc Lighting Company.

The lighting of Cleveland's Public Square, which required twelve 2,000-candlepower lamps placed on towers, was a demonstration intended to sell the Brush system to that city.[7] But the first Brush system actually sold had been purchased for domestic use by the wealthy Longworth family of Cincinnati. It was quickly discovered that theirs was an inappropriate application because of the intense glare and excessive heat emitted by the lamps. The first municipality to purchase and install a permanent arc lighting system in the United States was Wabash, Indiana. Four Brush arc lights were installed atop the Wabash County Court House, located on a hillside just north of the town's business district. A newspaper reporter observed the system's inauguration with great enthusiasm:

> At eight o'clock the ringing of the Court House bell announced that the exhibition was about to commence. Standing on the streets in front of the Plain Dealer Office, we hurriedly looked around to measure the general darkness as best we could. The city, to say the least, presented a gloomy, uninviting appearance. . . . Suddenly from the towering dome of the Court House burst a flood of light which, under ordinary circumstances, would have caused a shout of rejoicing from the thousands who had been crowding and jostling each other in the deep darkness of the evening. No shout, however, or token of joy disturbed the deep silence which suddenly settled upon the vast crowd. . . . The people, almost with bated breadth, stood overwhelmed with awe, as if in the presence of the supernatural.[8]

The reporter then moved around town to gauge the level of brightness. One street away from the courthouse he could read a newspaper held to the light; at two blocks away he could read headlines; at four blocks he could still discern the advertisements. The *Muncie Evening News* reported with some bias (from the heart of what was, after all, still Indiana's Gas Belt) that, contrary to rumors, it had not been possible to read a newspaper beyond the Wabash city limits.[9]

Charles Brush was one of the early heroes of electrical lighting and for a time stood shoulder to shoulder with the likes of Thomas Edison. So did Elihu Thomson and Edwin Houston, both of whom were teachers in Philadelphia when they began to experiment with electricity, first simply by placing some Brush lamps in a local store window. Then they designed an improved dynamo and arc lamp and installed them for a number of merchants. In 1880, Thomson joined an investor syndicate to manufacture arc lamps, dynamos, and related devices under Thomson and Houston patents (see fig. 2.1). Houston continued teaching high school, but he also kept experimenting, patenting new electrical welding, transformer, and metering devices.[10] The Thomson-Houston Electric Company would eventually merge with the Edison General Electric Company, forming today's General Electric. The Thomson-Houston management went on to control the consolidated firm.

FIGURE 2.1. The Thomson-Houston Electric Company's display at the Philadelphia Electrical Exhibition, 1883. Trade fairs held annually promoted the application of electricity, including its use in outdoor lighting. Shown here are the arc lamps of one of General Electric's founding companies. From *Electrical Review* 5 (Oct. 11, 1884): 1.

Numerous inventors experimented with arc lighting, many recruiting local investors to create manufacturing companies. Still other firms were formed around European patents brought to the American market. Many manufacturing companies, in turn, encouraged the formation of local electric light companies, some even operating central generating stations. The first such station, built in San Francisco in 1879, used Brush-licensed equipment.[11] Yet arc lighting actually represented a retreat from the large central-station idea pioneered by the gas utilities. Through the mid-1880s, most arc lighting systems involved very small generators serving only a few dozen lamps.

The early adoption of arc light, as with gas light, included theaters, saloons and beer halls, hotels, and department stores. The first hydroelectric plant in the United States was constructed in Minneapolis in 1882; it used four circuits to supply power to Brush electric arcs in saloons and retail stores from Bridge Square southeast along Washington Avenue paralleling the Mississippi River.[12] Only later were arc lamps used in street lighting in the same area. In Chicago, the adoption of arc light in the Academy of Music prompted a strike by actors, who were concerned that the intense brightness on stage would distort their appearance.[13] Very quickly arc lights became standard theater lighting, as arc lamps were also suspended from auditorium ceilings to light galleries and attached to outside marquees to light signs. Limelight projectors and gas footlights remained in use in many theaters until incandescent lighting came of age. Traveling circuses, most especially P. T. Barnum's, adopted arc lighting systems that, owing to their small scale, were easily transported from place to place. Most Americans outside the big cities saw their first arc lights in a circus tent. Large concert saloons, dance halls, and, of course, beer gardens replaced gas lighting with arc fixtures, the novelty of bright light serving to attract greater patronage.

Not only did the new lights attract, but in many instances they were also substantially less costly. Cost reduction proved to be the single most important factor in popularizing new electrical technologies. In 1880, the Grand Pacific Hotel in Chicago replaced 570 gas lights in its lobby and dining room with 17 arc lights, saving nearly $300 per month in lighting costs.[14] When insurance companies began to lower rates to customers who changed from gas to electricity, the savings grew even more. Chicago's first electric "central station" was established with a dynamo located in the basement of the local YMCA, which powered fifty arc lamps rented to merchants in the immediate vicinity. Such small systems quickly proliferated throughout Chicago's Loop, as the new light source brought prestige to businesses adopting it. "Burning the lights," as historian Harold Platt argued, became a conspicuous sign of affluence and luxurious living.[15] The well-to-do flocked to the light not only in theaters and hotels but also in their private clubs and exclusive restaurants.

The larger retailers from city to city embraced arc lighting quickly, especially the department store proprietors with facilities large enough to make installations economically practical. Charles Brush's first series system was installed in a Boston clothing store in 1878. That same year, John Wanamaker installed twenty Brush lamps in his Philadelphia establishment.[16] Merchants preferred the new electrical devices since, unlike gas fixtures, they did not give off the acidic fumes that discolored and deteriorated fabrics. Also, the new arc lamps did not produce the high levels of moisture which dampened store interiors and fogged show windows.

Excelsior Electric Co.

HOCHHAUSEN SYSTEM.

COMPLETE SYSTEM
——BOTH——
Arc and Incandescent
ELECTRIC LIGHTING.

Municipal Lighting
A SPECIALTY.

PERFECT •
AUTOMATIC
• **REGULATION**

Guaranteed to cut down to one Light, saving
power in proportion to the number
of lights burning.

Lights Perfectly Steady.
Free from Hissing and Flickering.

Catalogue and Full Information
FURNISHED ON APPLICATION.

EXCELSIOR ELECTRIC CO.
11 East Adams Street,
CHICAGO, - - ILL.

FIGURE 2.2. Advertisement for the Excelsior Electric Company of Chicago. Like most early
arc lamps, the glass globe shielded but did not seal the carbon electrodes, which were left
to glow brightly in the open air as electrical current arced between them. Advertising insert in Fred H. Whipple, *Municipal Lighting* (Detroit: privately printed, 1888), 236.

Arc lights quickly found industrial uses as well, especially in cotton and woolen mills. In 1879, one of the first Brush systems was placed in a worsted mill in Providence, Rhode Island, and by the end of year mills of Hartford, Connecticut, and Lowell, Massachusetts also had Brush or other systems.[17] Even in smaller cities arc light quickly found industrial application. In Indiana, Muncie's first electrical power company began when its promoter used arc lamps to light his flax-bagging factory, greatly reducing the fire hazard and, of course, his insurance costs.[18]

LIGHTING STREETS AND WATERFRONTS

Electric lighting debuted on the streets of New York City in 1880, and, as at Cleveland, it was in the form of a demonstration by Charles F. Brush. At his company's expense, twenty-three lamps were erected along Broadway, in a display that earned him city contracts to light Union and Madison Squares, both Broadway and Fifth Avenue between Fourteenth and Thirty-fourth Streets, and Fourteenth and Thirty-fourth Streets between Broadway and Fifth Avenue. Towers were erected in the squares, but lampposts were used along the streets (see fig. 2.3). By 1886, New York City had thirty miles of thoroughfares lit by arc lamps placed at 250-foot intervals.[19] However, arc lighting on Fifth Avenue, the city's prime residential show street, was dismantled after residents objected to the unsightly wires connecting the fixtures. High rents for underground conduits still made arc lighting with buried cables too expensive and, through the 1880s, only a few private systems operated on Fifth Avenue as the street "returned to the gloom of gas."[20] (Fifth Avenue eventually would be lit by incandescent lamps attached to Edison's commercial power grid.) By 1893, New York City had in operation on its streets 1,535 electric arc lights, as opposed to 26,524 gas lights.[21] By 1913, the number of arc lights had climbed to about 19,250, as opposed to about 18,000 incandescent lamps and about 46,500 gas lamps of various descriptions.[22]

Chicago's arc light illumination also began in 1878 with a demonstration, the handiwork of John P. Barrett, who, after the city's disastrous fire, had installed the city's sophisticated telegraph fire alarm system. The two battery-powered lamps mounted on the city's northside water tower elicited sarcastic comment from the *Chicago Tribune:* "The gleam slowly

FIGURE 2.3. Electric arc lights in New York City's Madison Square, 1881. Lamps were positioned on towers high above the square as well as on posts marginal to its sidewalks. Elevating arc lamps above the line of sight greatly reduced the problem of intensive glare. As with gas lighting, New York City's business streets were the first lit with electricity, starting with Broadway. From "The Electric Light on Madison Square, New York," drawn by Charles Graham, *Harper's Weekly*, Jan. 14, 1882, 25.

gathered into a ray, which slanted through the darkness, and succeeded in bringing into unpleasant prominence a cow shed in the vicinity. The ray at its brightest was about a foot broad at the start, and widened to perhaps twenty feet at the base on the ground."[23] The light Barrett installed was unsteady, suffering from constant flicker. Public lighting with arc lamps in Chicago would not begin in earnest until 1887, and then only on Chicago River bridges. An investment syndicate formed as the Chicago Arc Light and Power Company consolidated a host of small firms supplying private customers. The new firm was driven by the promise of lucrative municipal contracts for street lighting, which had previously gone to local gas interests. By 1910, arc lamps were widely distributed along Chicago's principal streets.

Waterfront illumination like that in Chicago helped encourage public arc lighting in other cities, like New Orleans. In 1881, the New Orleans Brush Lighting Company, organized with local capital, placed forty lights in the large Crescent Hall, followed a year later by a hundred 2,000-candlepower lamps along five miles of wharf and riverfront.[24] A separate generating plant was then constructed for the lighting of roughly three miles of Canal Street and then of Royal and Chartres Streets in the Vieux Carre. Of New Orleans' 655 arc lights in 1885, 450 were subsidized through private subscription, and the remaining 205 were wholly owned and operated by the city. It is clear that the initial adoption of electric street lighting in the United States varied substantially from city to city, as a function of local politics, the felt need for the new technology (itself a function of already existent gas lighting), and the preferences of lighting enthusiasts.

TOWERS

Powerful arc lamps were developed that could deliver light up to 4,000-candlepower. Such lamps, clustered and fixed at great heights, were capable of lighting whole city neighborhoods. The lighting installation at Wabash, Indiana, was one of the testing grounds for such ideas. Indeed, as early as 1802 one Benjamin Henfrey had erected in Richmond, Virginia, an oil "thermolamp" on a tall column, intended to flood the city with light. Unfortunately, the experimental oil lamp cast far less light than was hoped for, but grand ideas continued to arise.[25] James Silk Buckingham, the English author and traveler who had described New York City's gas lights as lacking, proposed for Britain a utopian city to be called Victoria. At its center was to be a 250-foot light tower equipped with arc lamps.[26] A 1,200-foot "sun tower" was proposed for the center of Paris, to be located near the Pont Neuf, in anticipation of the 1889 Paris Exposition. A. G. Eiffel's tower was erected instead, but it was never decked with arc lamps. Lesser projects undertaken in both European and American cities demonstrated the ineffectiveness of light towers in "turning night into day."

In the United States, planners in Washington, D.C., considered tower lights for the Washington Monument as it neared completion, with the idea that the monument could serve as a light platform. Test lamps were placed on the Smithsonian Institution and on the Capitol, and all gas lamps were extinguished on the surrounding grounds and streets.[27] The experiment did not excite admiration, however, and arc lighting was rejected. The idea behind such floodlighting was simple. From a high platform, a flood of light would enable pedestrians, those in wagons or on horseback, and others to distinguish objects silhouetted in the streets (see fig. 2.4). Lighting technicians did not pretend to replicate daylight so much as moonlight. Numerous cities across the United States adopted tower lighting, including Akron, Buffalo, Chattanooga, Denver, Evansville, Fort Wayne, Kansas City, Los Angeles, Louisville, Minneapolis, Mobile, San Francisco, and San Jose. Lighting towers survive today in Austin, Texas (fig. 2.5).

Many cities erected only one or two towers and then turned to traditional lighting with lampposts. In 1883, Minneapolis constructed an "electric moon" comprising eight arc lamps on a 257-foot tower in Bridge Square, then the center of the city's business district.[28] Other cities embraced light masts in downtown districts, using other street lighting in peripheral areas. Los Angeles erected thirty-six towers—fifteen of them 150 feet tall with three lamps of 3,000 candlepower each, the remainder sixty feet tall with weaker single lamps.[29] But only one major American city adopted towers for all of its lighting needs. That city was Detroit.

In Detroit, 122 towers were erected to illuminate some twenty-one square miles of city space. All towers at the center were 175 feet high, carrying six 2,000-candlepower lamps each. Away from the center, 150-foot

FIGURE 2.4. Advertisement for the Star Iron Tower Company of Fort Wayne, Indiana. Arc lamps on tall towers cast a faint "moonlight" sufficient to enable pedestrians and those in carriages and wagons to see obstructions silhouetted in city streets. From *Electrical Review* 13 (Feb. 10, 1889): 36.

FIGURE 2.5. Light Tower at
Austin, Texas, 1996. This is
one of several light towers
preserved in Austin as historical landmarks.

masts held four lamps each. The towers, most of them located at street intersections, stood on single legs held upright by cables (see fig. 2.6). In
most other cities, lights on towers were lowered to the ground by cable
and pulley for maintenance. But in Detroit, trimmers raised themselves to
the tops of the towers daily by man-powered elevators balanced on
weights. Whereas traditional street lighting could vary in intensity from
street to street according to demand, tower lighting created uniform car-
pets of light that bore little if any relationship to traffic and other needs
below. As one observer commented: "A twilight glow was shed over a
wide area, but there was no effective lighting anywhere."[30] The technology was more spectacular than it was effective.

FIGURE 2.6. Postcard View of Detroit's Campus Martius, circa 1900. This was the ceremonial heart of the city, from which major thoroughfares radiated. A light tower stands in front of city hall, marking the apex of city mass transit. The coming of the automobile, however, quickly made the city's tower lighting obsolete.

Detroit located the towers in its business district 1,000 to 1,200 feet apart, compared with 2,500 feet in the residential areas beyond. Even so, the buildup of nighttime traffic at the center quickly necessitated a complementary system of electric lamps on posts. In the city's residential zones, thick tree canopies (Detroit was then known as the "City of Trees") made tower lighting ineffectual in the summer months. After five years, Detroit began to dismantle its towers; only those on Cadillac Square, the city's ceremonial heart, survived to the turn of the twentieth century. Within a few years, they too were removed. As historian Wolfgang Schivelbusch observed, technical rationality simply overshot the mark, ending up as "technical fantasy and pipe dream."[31] Here was a technical solution to a city's street lighting that seemed at first to offer efficiency. It was, after all, cheaper to service a hundred towers than to service thousands if not tens of thousands of individual street lamps. What was created, however, proved merely an exercise in technological monumentalism.

A few small towns adopted towers with arc lights. Elgin, Illinois, erected twenty-three lamps across seven towers. A reporter waxed romantically upon the system's inauguration: "The fronts of the blocks were brightened by the halo; the fulgor [sic] stole across the black river and made silvery pathways; the very air seemed warmed by the gentle influence; and when we stood where we could see all the towers at a distance, the lights appeared merged into one large ball on each structure, and to stand like sentinels at convenient intervals, watching over the destinies of a busy city."[32] It was a "day of jubilee," for the "Queen City of the North-west" was now "radiant twenty-four hours a day." But even in small places like Elgin the promise of lighting whole landscapes from towers proved short-lived. Yet tower lighting has enjoyed some revival in recent decades. Today, masts of over 100 feet high hold clusters of modern vapor lights at freeway interchanges, a practice started in the 1970s in the states of Texas, South Dakota, and Washington.[33]

LIMITATIONS

Although early arc lamps gave more light per dollar invested than any other previous illuminant, they came with their problems. They required daily "trimming," the process by which rods were replaced and lamps adjusted. Arc light was exceedingly brilliant, glowing with a sharp greenish yellow, and the brilliance could not be "subdivided"—that is, reduced in intensity and thereby made more subtle. The exceptionally high voltage made it difficult to turn individual lamps on and off without affecting the efficiency of lamps remaining in a series. Open carbon arc fixtures quivered as they went from full, half, and quarter light in a regular cycle of pulsation. This flickering was both an annoyance and a benefit to visibility, as it tended to sharpen the silhouetting of objects against dark backgrounds. When alternating current was used, the open air lamps produced humming noises both in the lamp mechanism itself and in the air between electrodes. Arc light was clearly unacceptable for interior lighting except in the largest of auditoriums, not only because of its glare but because of the excess of heat released.

Arc lamp technology improved substantially in the 1890s, keeping it apace of improving gas mantles and, even more significantly, the rapidly

FIGURE 2.7. Advertisement for the Warner Arc Lamp Company. Arc lamp technology culminated with enclosed fixtures. As with the gas mantles, glass surrounds sealed the lamp in a vacuum, greatly extending lamp efficiency and lowering cost. From *Electrical Record* 2 (Dec. 1907): 109.

evolving incandescent filament lamp. In turn, the enclosed arc, open flaming arc, enclosed flaming arc, and magnetite luminous arc lamps came into more regular use. In the enclosed arc lamp (see fig. 2.7), as the name suggests, a glass globe surrounded the arc to restrict air intake. With reduced combustion, rods lasted up to 150 hours, vastly reducing the need for trimming and other maintenance costs. These lamps also produced a steadier light, although at the cost of reduced luminance. Through the burning of evaporated salts in the flaming arc, the light color problem was solved, as the harsh greenish-yellow glow could be turned into rose tints more flattering to people. Short life-span remained a problem, which was only partly solved by the enclosed flaming arc lamp, a modification that increased both operating life and luminosity. A tightly fitted globe restricted the access of oxygen to the arc, and a large metallic chamber placed just above the arc received and condensed the metallic oxides, which would otherwise condense on the globe and obstruct the light.[34]

The most innovative lamp, the magnetite or luminous arc lamp, extended electric arc lighting well into the twentieth century. The upper electrode, or anode, consisted of a rod of solid copper resistant to burning. The lower electrode was composed of various metallic oxides packed into an iron tube. As in the case of the flaming arc lamps, it was, indeed, the arc stream, and not the tips of the electrodes, that furnished the light. Magnetite was a good conductor of electricity and was rapidly vaporized in the arc, which made the flame large, while the titanium made it very luminous.[35] The result was a brighter, better balanced (although somewhat purplish), and steadier light. More importantly, the intervals between trimmings could be increased up to 350 hours. Power company or municipal street lighting trucks equipped with extension ladders facilitated the cyclical work of trimming the lamps (fig. 2.8).

The introduction of arc lighting made good on the often-used metaphorical description of streetlights as artificial suns.[36] The arc lamp, in its various forms, was indeed an artificial sun, as the spectrum of light cast was similar to that of sunlight. Under the intense arc light, the eye could see with the retinal cones, as it did during the day, whereas, with the dim gas light, perception took place more with the retinal rods. Likewise, staring directly at an arc was painful to the eyes. Charles Brush mused late in life about the "tedious education of the public to the new light."[37] At Cleveland, in that first street lighting demonstration in 1879, many people brought colored spectacles or pieces of smoked glass, determined to look directly at the arcing electrodes. The lamps were not to be stared at, Brush argued. What counted was the light cast and reflected as illumination. It is clear, however, that few people were able to overlook the bright glare. And in some instances that glare could be disorienting and dangerous. An electric arc tower at Hell Gate on the East River in New York City was removed after pilots in the harbor reported that the glare was so intense it made keeping to the narrow channel impossible at night.[38]

An arc light mounted on a standard lamppost created a brilliant pool of light immediately below, luminance falling off in all directions. To a

FIGURE 2.8. Trimming an arc lamp on a Manhattan street. Labor costs associated with the frequent trimming of electrodes, in addition to the glare that arc lamps emitted, led to the general abandonment of arc lamp technology by 1930. From *Thirty Years of New York, 1882–1912* (New York: New York Edison Co., 1913), 199.

pedestrian, city streets still appeared as separated pools of light, just as with gas lighting. Only now these bright places were larger and more intense. When lamps were spaced close together, pools of light merged. Within each light pool the arc shined with a concentrated brilliance. Things under this umbrella reflected light sharply. As one observer noted:

"Every minute globule of water which floats in the air, every particle of dust, every microscopic insect and seed and spore, intercepts a portion of [the] beams, and becomes a new center of radiation."[39] At a distance the light was more diffuse, neutralizing shadows and thus giving the effect of moonlight. As one English visitor to New York City observed in 1883: "Looking along Broadway, New York, one sees the bright, white, moonlight effect of electric lamps every here and there—here in front of a store, there in front of a theater, beyond in front of a hotel."[40]

It was the tower lighting, however, that excited the greatest imagination at first. Here the new lighting was employed in a very novel fashion worthy of comment. Our English visitor also described Madison and Union Squares:

> The effect of the light in the squares of the Empire City can scarcely be described, so weird and so beautiful is it. Enormous standards, rising far above the trees, are erected in the center of each square. From these standards the light is thrown down upon the trees in such a way as to give them a fairy-like aspect. Except for the temperature, it would be easy to imagine, even on summer nights, that they were covered with hoar frost. Immediately beneath the standards the shadow of every leaf and branch of the interposing trees is imprinted on the asphalte [sic].[41]

Trivial as it may seem to us now, never before had such a thing been seen, and at the time it was something to marvel at.

Electric lighting became a new symbol of the emergent modern world, a role the whole of electrical technology quickly assumed. Brighter came to stand for better, a meaning the adjectives *bigger, stronger,* and *faster* also conveyed. Writing in one of the early trade journals of the electrical industry, an editor speculated on lighting in Chicago's future. "We hope the day is not far distant when the brilliant and penetrating arc-lamp will supersede the sickly gas-lamps which now vainly struggle to illuminate Chicago's magnificent boulevards. It is time the pallid gas jets were laid away in innocuous desuetude to make place for their incomparable successor, electricity."[42]

As electricity and electrical lighting entered American consciousness, so also did the new technology impact American lingo. American English quickly became saturated with relevant expressions. "An 'energetic' person was 'a human dynamo,' a powerful performer was 'electrifying,' and an angry person might 'blow a fuse.'" Americans could be "switched on," "turned off," "overloaded," "shocked," and "short-circuited." They could "see the light" and have "bright ideas."[43] As metaphor, things electrical were associated with intelligence and drive. Above all, electric light meant progress. Its adoption was seen as progressive.

ALTHOUGH IT WOULD BE REPLACED by more sophisticated electrical illumination, arc lighting held its own for decades and could still be found in several American cities as late as World War II. Indeed, the arc lamp, as a "discharge" device, carried over into sodium vapor, mercury vapor, and other forms of lighting widely adopted after 1930. Electric "arcing" through one gas medium or another characterizes most street lighting today. In the late nineteenth century, arc lamps, augmented by residual oil and gas fixtures, brought to American cities unique mixtures of light. The American night was made visually complex in its illumination because of the various overlapping technologies. Flickering oil and gas lamps casting their small pools of light on streets remained a part of the American urban night long after the arrival of electric arc lighting. The new electric light was quickly applied to illuminating the most heavily traveled streets, especially in downtown districts.

Electric arc light cast from towers proved too diffuse and too dim to accommodate the automobile, another great symbol of the dawning modern age. Detroit, the "motor city" by 1920, discarded its light towers soon after their construction. Visual effectiveness was now what counted, and arc lamps on towers served the needs of motorists poorly. However, arc lighting on poles along streets at ground level proved problematical because of their intense glare. Arc lighting provided too much of a good thing, a problem only partially solved through the use of shielding luminaires. The application of incandescent filament lamp technology soon filled the need for bright light that could be managed, and, thereby, made comfortable to the eyes. During the brief period when arc lamps were clustered high above city streets on towers, streets were sustained in lightened shadow, but it was shadow nonetheless. The visual effect of moonlight preserved the traditional urban night of contrasting light and dark, in which diverse kinds of illumination competed successfully for the eye's attention. Once arc lamps were placed close to street level, however, a very different nighttime city emerged. Here was night-destroying brilliance that suffered from excess. It did, however, promise much for the future: not the romantic nighttime city or the city of nighttime danger but instead the city of nighttime functionality. The light only had to be better articulated.

CHAPTER THREE

Maturation of the Lighting Industry

Incandescence refers to the glow produced by an object at high temperatures, that is, the emission of radiation rendering it visible as a glowing object. With lamps, incandescence is best achieved within the evacuated seal of a glass container. Incandescent light quickly took on a rather narrow definition having to do with the kind of standardized electric lamp, or bulb, created to produce it. An incandescent electric lamp consists of a filament in a glass globe filled with inert gas, today usually a mixture of nitrogen and argon. Its major parts are the filament, the bulb surround, the gas fill, and the base (fig. 3.1). Incandescent light is produced when electric current passes through the filament, causing it to glow. The new electric lamp proved revolutionary, for it solved the problem of the arc light's indivisibility. Incandescent bulbs could be made to light at various levels of intensity, and they could be used interchangeably, one lamp fixture to another. Immediately it seemed to be a form of electric light ideally suited to domestic use indoors. Later, it would be adapted to outdoor applications.

THOMAS EDISON

Thomas Edison's name has become synonymous with incandescent lighting and with electric lighting generally. Many Americans, who have forgotten or never knew about arc lighting, assume that Edison alone was responsible for bringing electric lighting to the fore. The invention of practical incandescent lamps belongs as much to Joseph Swan of England, who, working independently, developed at about the same time as Edison

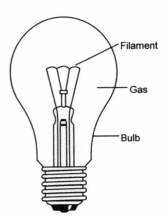

FIGURE 3.1. Basic elements of an electric incandescent lamp. Light is produced by the glow of an electrically excited filament enclosed in a vacuum to sustain filament life. Modified from *Roadway Lighting Handbook* (Washington, D.C.: Federal Highway Administration, 1978), 37.

a workable incandescent electric lamp. Heinrich Gobel, a German emigre to America, perfected an incandescent lamp with carbon filament as early as the 1850s, but its very short life-span made it impractical as anything but a curiosity.[1] Edison triumphed on several fronts. He perfected a long-lasting filament out of carbonized bamboo, a filament material that even Swan acknowledged as far superior to anything previously known. Edison's truly ingenious contribution was to conceptualize incandescent lighting as a total illumination system. His new lamp was but one part of that system. Subsequently, Edison's much-publicized patent defense of his lighting system established him in the popular mind as the sole developer of electric incandescent lighting (fig. 3.2).

Edison's light bulb evolved by stages. At first he experimented with platinum spun into strands, as others had done before. He called these thin wires *filaments*, after the Latin word *felare*, which meant "to spin."[2] There began a long process of trial-and-error, whereby other metals and then various carbonized plant fibers were tried and discarded. Finally, carbonized filaments of one specific species of bamboo yielded the desired long-lived incandescence. A good part of this success was attributable to a perfected method for sealing vacuums in glass bulbs, an endeavor as lengthy and as frustrating as Edison's search for a filament. Edison had already made himself a reputation with the phonograph, the telegraphic stock ticker, and the telephone transmitter, among other inventions. With earnings from these devices he had created a research laboratory at Menlo Park, New Jersey, one of America's first "research and development" fa-

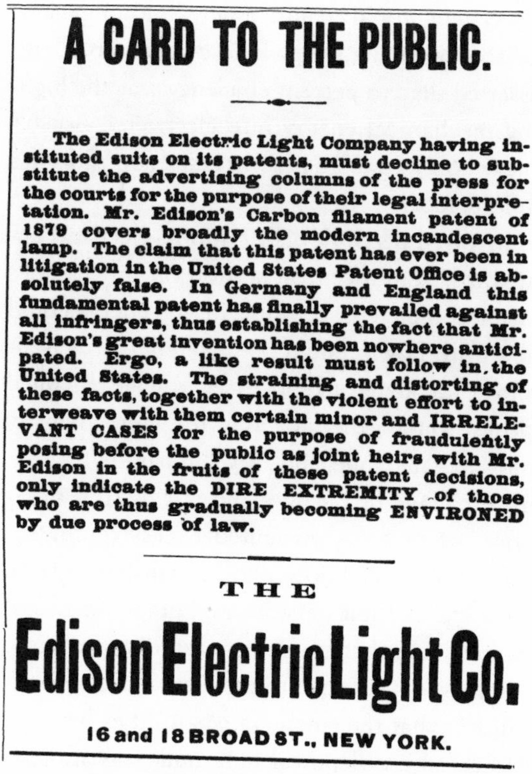

A CARD TO THE PUBLIC.

The Edison Electric Light Company having instituted suits on its patents, must decline to substitute the advertising columns of the press for the courts for the purpose of their legal interpretation. Mr. Edison's Carbon filament patent of 1879 covers broadly the modern incandescent lamp. The claim that this patent has ever been in litigation in the United States Patent Office is absolutely false. In Germany and England this fundamental patent has finally prevailed against all infringers, thus establishing the fact that Mr. Edison's great invention has been nowhere anticipated. Ergo, a like result must follow in the United States. The straining and distorting of these facts, together with the violent effort to interweave with them certain minor and IRRELEVANT CASES for the purpose of fraudulently posing before the public as joint heirs with Mr. Edison in the fruits of these patent decisions, only indicate the DIRE EXTREMITY of those who are thus gradually becoming ENVIRONED by due process of law.

THE

Edison Electric Light Co.

16 and 18 BROAD ST., NEW YORK.

FIGURE 3.2. Advertisement for the Edison Electric Light Company. Edison's company spent a decade in federal courts establishing its patent rights to his central system, Thomas Edison's real contribution to electric lighting. The improved incandescent bulb was but one part of that system. From *Electrical Review* 10 (May 28, 1887): 15.

cilities dedicated to innovation through experimentation. With a workforce of more than one hundred technicians, Edison pursued the idea of incandescent lighting across several fronts.

Edison succinctly put forth his intentions in one of his notebooks. "Object, Edison to effect imitation of all done by gas, so as to replace lighting by gas, by lighting by electricity."[3] He had made a thorough study of gas illumination and was convinced that what was needed was an analogous electrical generating and distribution system. Edison's focus on incandescence using filaments contained in vacuums followed from his understanding of Ohm's law, formulated in 1827 by G. S. Ohm, which stated that electrical current flowing through an electrical circuit was inversely proportional to the resistance of the current. By increasing filament resistance, the current could be kept relatively weak. Edison's, therefore, would be an electrical distribution system of low, constant voltage. He knew that, if perfected, this system would require a new kind of genera-tor, and thus Edison turned to developing a new kind of dynamo. Com-mon wisdom at the time asserted that 50 percent efficiency was the high-est attainable in converting mechanical energy into electricity. Brush's dynamos, for example, achieved only 30 to 40 percent efficiency.[4] Edison's new machine would reach 90 percent.

Edison's lighting system involved network distribution—electrical en-ergy being transmitted outward from a central station on power lines forming a grid. Networks were completely accessible in that energy could be tapped at any point on a grid. Power could be metered as it was si-phoned away, allowing cost to be accounted and prices set. Voltage remained constant. That is, lamps closest to the generator lit at the same intensity as those farthest away. Lamps operated independently, "in parallel" rather than "in series," and lamp breakage or burnout did not disrupt other lamps. Circuit breakers and safety fuses prevented excessive current buildup, whereas switching gears enabled power to be transferred from network to network, interlinking multiple generators. Lamps (or bulbs) were standardized, making them interchangeable between different lamp fixtures. Each part of the system was engineered to support all other parts.[5]

Edison's system did replicate what the emergent gas utilities had cre-ated. Indeed, Edison's standard 18-candlepower light bulb was adopted to simulate the luminance of the standard domestic gas jet. But Edison's system was better than anything gas could offer. It was cleaner in that it did not leave a gritty residue on furnishings. It did not deteriorate fabrics. It carried no risk of explosion and relatively little risk of fire. Edison's elec-tric light did not generate moisture indoors, and it did not consume oxy-gen, causing headaches and nausea. And, as the new electrical technology matured, it proved cheaper than gas for domestic lighting and, eventually, for street lighting as well.

The first public demonstration of Edison's system took place at Menlo Park in December 1879. Two lights framed the gate leading to the office building, eight were located on poles outside the laboratory, and thirty were fixed inside. Another two dozen lamps were strung along the street in the direction of the nearby railroad depot. One visitor that first night was Henry Villard, organizer of German investor syndicates in North America and president of a steamship line. The first commercial installation of the new Edison technology, therefore, would be made aboard the steamship S.S. *Columbia*. Another visitor that night was William Sawyer, developer of a less satisfactory incandescent lamp, who would, with others, consume much of Edison's time and energy in court litigation contesting patent rights.

Thomas Edison was always the self-promoter, rarely missing an opportunity to publicize his accomplishments. His mythic image as "the Wizard of Menlo Park" was amplified over the years by corporate publicity departments. His praises were literally sung in corporate promotion:

> *Let there be light*
> *The Wizard cried,*
> *And straight the night*
> *Was glorified,*
> *While arc and incandescent blazed*
> *Till all the world looked on, amazed*
> *And dazzled by the splendid light*
> *Which swept the shadows of the night*
> *Away*
> *And turned the darkness into day;*
> *Lit up the city*
> *Flashed its gleams*
> *Along the pathways of man's dreams*
> *Of hidden power, that he might see*
> *The trail to untold energy.*[6]

Edison became an American legend. Henry Ford, another legend, later moved sections of Edison's Menlo Park laboratory to his outdoor history park at Greenfield Village in Dearborn, near Detroit. There, on the fiftieth anniversary of Edison's discovery of his filament, Ford, Edison, and President Herbert Hoover gathered to celebrate, their ceremony broadcast to the nation over radio. Long had Edison been the celebrity. A visitor to New York City early in the twentieth century saw Edison in a restaurant and later wrote, "Electricity is the true 'white magic' of the future; and here, with his pallid face and silver hair, sat the master magician—one of the great light-givers of the world."[7]

CORPORATE STRUCTURING

The Edison Electric Light Company was organized with capital supplied by Hamilton Twombly, president of the Western Union Telegraph Company, Twombly's father-in-law, W. H. Vanderbilt, and financier J. Pierpont Morgan. Edison took 2,500 of the original 3,000 shares, most of which he subsequently sold in establishing operating subsidiaries that preserved the parent firm's original purpose of research and development. One such satellite company was the Edison Electric Illuminating Company of New York, the operator of Edison's first central generating plant on Pearl Street in New York City. In a host of cities across the United States there were other power utilities wholly or partly owned by the parent firm. Another subsidiary, the Edison Company for Isolated Lighting, manufactured and installed small lighting systems. The Edison Machine Works manufactured powerplant machinery, including the "Jumbo" dynamos installed at Pearl Street. The Electric Tube Company and the Edison Lamp Works were other Edison firms.

In New York City, Edison's central system service was initiated in 1882, extending first to buildings close to the Pearl Street station. The offices of the *New York Herald* were lit, as well as the Barnes, Polhemus, and Drexel Buildings, the last housing the headquarters of Drexel, Morgan, and Company. As the first electric lines led up Fulton and Nassau Streets, stores located there were lit as well. The *New York Times* Building was outside the first lighting district and so was equipped with a small, isolated system. A reporter described the new technology:

> The whole lamp looks so much like a gas burner surmounted by a shade that nine people out of ten would not have known the rooms were lighted by electricity except that the light was more brilliant than gas and a hundred times steadier. To turn on the light nothing is required but to turn the thumb-screw, no matches are needed, no patent appliances. As soon as it is dark enough to need artificial light, you turn the thumb-screw and the light is there. No nauseous smell, no flicker, no glare.[8]

Edison was not the only one installing incandescent lighting. Also at work in New York City was the United States Electric Lighting Company, which in 1880, just after Edison's equipping of the steamship *Columbia*, installed a lighting system in the offices of the Safe Deposit Company (fig. 3.3).[9] Edison's first isolated system on land was installed in a Manhattan lithography shop in 1881, and other systems followed quickly in offices and retail stores. By June 1882, the Edison Isolated Lighting Company had installed some sixty-seven plants across the United States supplying more than 10,000 lamps.[10] By 1885, the number of customers had grown to 494 plants, with more than 181,000 lamps in service.[11]

Edison's great preoccupation was promoting his central system approach. The Pearl Street powerhouse, not the many small, isolated generators, represented the future. Originally, the station was a four-story rebuilt warehouse which fit into the street rather than standing apart like later powerplants. Divided by a firewall down the center, the plant generated power on one side and stored the tubes, wires, and other materials used in laying underground conduits on the other side. The six generators were nicknamed "jumbos" by the media owing to their 125-horsepower capacity and enormous size—their armatures alone weighed some six tons each.[12] The Pearl Street Station was the largest that Edison had yet undertaken, but it was truly a small-scale enterprise by later standards, employing, as it did, only several dozen employees (fig. 3.4). By 1930, the company employed more than 36,000 workers and had a half dozen generating plants on line, including the Hell Gate Station, producing 810,000 horsepower.[13]

Edison encouraged the elimination of unsightly poles and wires from city streets. In most big city downtown areas, masses of wires were strung along street margins, often obscuring building facades. They inhibited firefighting and were themselves considered a fire threat by critics. Edison instead placed his company's wires in underground conduits, a practice soon followed throughout most of lower Manhattan (fig. 3.5).

Recognizing electric incandescent lighting as a likely successor to gas illumination, in 1900 New York City's Consolidated Gas Company began purchasing local electrical companies, including the New York Gas and Electric Light, Heat, and Power Company. It then consolidated with the Edison Illuminating Company a year later to form New York Edison. More purchases followed, and the amalgam of utilities finally merged together in 1936 as Consolidated Edison.[14] Most central systems were small

FIGURE 3.3. Advertisement for the United States Electric Lighting Company. Although first to actually illuminate buildings in New York City with the new incandescence, this company failed to replicate Edison's central system technology and could not, accordingly, achieve the scale efficiencies necessary to remain competitive. From *Electrical Review* 10 (Aug. 27, 1887): 27.

FIGURE 3.4. Edison Illluminating Company's Pearl Street Generating Plant, 1882. From this primitive beginning in an old warehouse, the Pearl Street facility was rapidly enlarged into a modern power generating plant. From Charles T. Child, "Twenty Years of Electrical Development," *Electrical Review* 40 (Feb. 15, 1902): 201.

by later standards, having been designed only to supply, through municipal franchise, urban lighting districts limited in scale. When Samuel Insull, previously Thomas Edison's private secretary, traveled to Chicago in 1892 to assume control of the Chicago Edison Company, that firm was but one of twenty small electric lighting utilities in the city. Small-scale technology dictated corporate diversity in Chicago, as did the politics of franchising public rights-of-way separately for conduits and lines.[15]

A group of Chicago businessmen had formed the Western Edison Light Company in 1882, primarily to serve residential customers. In 1887, the company reorganized as Chicago Edison and, under Insull's leadership, bought its largest competitor, the Chicago Arc Light and Power Company, in 1893. Arc and incandescent lighting operated on different circuits with different voltages, and the two types of lamps could not be operated directly from the same generator. But, Chicago Edison found it could install storage batteries to be charged during the day and use them to power incandescent lamps at night.[16] In 1898, another Chicago firm arose, called

FIGURE 3.5. The Johnstone System of Underground Conduits. Edison was not alone in promoting the burial of power lines, following the example of gas mains. Note that in this 1887 engraving the street lights are arc lamps. An incandescent bulb is only pictured underground. Electric incandescent lamps were used through the 1890s almost exclusively for indoor lighting. From *Electrical Review* 13 (Feb. 23, 1889): 9.

Commonwealth Electric, and bought additional power companies operating in the Chicago area. In 1907, Chicago Edison and Commonwealth Electric were merged as Commonwealth Edison, now ComEd.

Samuel Insull's drive to consolidate Chicago's electric utilities hinged on building larger and larger generating plants. Rates could be lowered through economies of scale, and disadvantaged competitors could be encouraged into the corporate fold. Insull's acquisition of the large Harrison Street powerplant, opened in 1894, was key to attaining quick centralized control in downtown Chicago. Included in the takeover were eighteen central stations and 498 self-contained systems supplying about 273,000 incandescent lamps and about 16,400 arc lamps.[17]

Insull emphasized innovation across his system of power generation and distribution. An improvement in one sector necessarily drove improvement in another. Besides seeking economy of scale by using large generators (both steam and water turbine), he massed generating units near load centers or market demand, economical sources of energy, and sources of water for cooling. He transmitted electricity to load centers using high-voltage transmission lines. Disregarding Thomas Edison's anathema for alternating current technology, he used it to transmit power over long distances, converting it in transformers back into direct current for local distribution. Insull cultivated mass consumption by charging low and differential rates, allowing supply to create demand. This demand, evened out over the diurnal cycle, enabled the company to use its generators more efficiently. He interconnected power plants in large grids in order to optimize their different characteristics, and he used load forecasting to achieve optimum effectiveness across this interconnected system. Insull also accepted government regulation in the establishment of a "natural monopoly" and settled for relatively low but steady returns on investments in order to obtain capital for expansion at reasonable interest rates.[18] According to historian Harold Platt, such light and power systems of America were neither inevitable nor driven by an inherent technological imperative.[19] Personalities like Samuel Insull, their business decisions, related public policies, and, finally, consumer preferences drove the industry, not the "inexorable logic of the machine."

GENERAL ELECTRIC AND WESTINGHOUSE

The parent Edison Electric Light Company merged in 1889 with its various manufacturing subsidiaries to form the Edison General Electric Company. Manufacturing and electrical utility functions could now be separated and, more importantly, Edison's sense of system could now be applied in management as well as in technological innovation. Brought into the consolidation were other firms including the Sprague Electric Railway and Motor Company. The new corporation was controlled not by Thomas Edison but by Henry Villard and Werner Siemens of the German Siemens and Halske Company. For years Edison had been only a minority stockholder, although he remained a director of the firm and head of research and development; Edison's persona as a technical genius was still important in promoting the reorganized company. At the heart of Villard's endeavor was a scheme to form an international cartel that could control the electrical industry worldwide.[20]

Henry Villard was, like Edison and Insull, one of the key figures directing electrical technology's application. After having emigrated to the United States in 1853, he struck up a business relationship with financier J. P. Morgan, using German capital to organize for Morgan not only a steamship line but also a western railroad empire that included both the Union Pacific and Northern Pacific Railroads. Villard also bought control of the New York *Evening Post* and took an active role in the affairs of the early Edison Electric Light Company, serving as a member of its board. He was responsible for the adoption of Edison technology in Germany—indeed, German Edison at Berlin was the most successful of the early Edison licensees.[21]

Edison General Electric's principal competitors in the United States were the Westinghouse Electric and Manufacturing Company (formerly the Union Switch and Signal Company) and Thomson-Houston. There was much cross-licensing among a wide diversity of firms as the industry sought to standardize electrical technology. Cross-licensing invited acquisition, merger, and consolidation so as to reinforce and strengthen patent rights. Edison General Electric and Westinghouse entered merger talks, but in 1896 the two firms settled for a mutual-licensing agreement. George Westinghouse (the developer of the air brake) had redirected his railroad equipment company toward the manufacture of electrical generators and incandescent lamps. Westinghouse was as prolific an inventor as Edison, certainly rivaling Edison's breadth of interests, but Westinghouse lacked much of Edison's zeal for publicity and self-promotion.

What pushed the Westinghouse Company to prominence was its embracing alternating current technology. As the name implies, a generator creates a flow of electricity that rapidly alternates forward and backward in a circuit. Direct current, in contrast, flows in a single direction around a circuit. High-voltage alternating current could be sent long distances efficiently in inexpensive, thin copper wires. Direct current required expensive, heavy cabling, and voltage levels fell very quickly with distance. French inventor Lucien Gaulard patented a transformer that could modify the high-voltage current of an alternating-current generator for low-voltage, direct-current circuits. Westinghouse bought the American rights.[22] The first alternating-current lighting system was installed at Buffalo, and, by 1889, systems perfected by Westinghouse engineers supplied more than 300,000 lamps nationwide. Cities adopting the technology included Baltimore, Denver, Minneapolis, New Orleans, Pittsburgh, Richmond, St. Louis, and even Schenectady, New York, the home of General Electric's principal research facility.[23] Westinghouse absorbed dozens of electrical companies, large and small, including the United States Electric Light Company.

The Edison General Electric Company consolidated with Thomson-Houston in 1892 to create the General Electric Company. Thomson-Houston had spent little money on developing new technologies; instead, the firm purchased patents and, according to critics, also infringed on the patents of others. It had, for example, bought the Brush Electric Company, and all of its arc lighting rights in 1880. Thomson-Houston's founder and president, Charles A. Coffin, was an accomplished businessman whose firm was financially solid. Villard's firm, on the other hand, suffered from large loan obligations, especially to J. P. Morgan, and Edison, for his part, was in the process of leaving in a buyout, many of his original patents about to expire. Thus it was Charles Coffin who assumed company leadership.[24] The new General Electric controlled approximately 75 percent of the American electrical lamp market, with two huge incandescent lamp factories, the former Thomson-Houston works at Lynn, Mas-

FIGURE 3.6. Advertisement for the Westinghouse Electric Company. Thomas Edison's stubborn adherence to direct-current technology opened the market to George Westinghouse and his alternating-current technology. Power could be transported longer distances and at less cost with alternating current. From *Electrical Review* 11 (Feb. 18, 1888): n.p.

sachusetts, and the former Edison General Electric works at Harrison, New Jersey. The company employed more than 10,000 people and had its headquarters at 44 Broad Street in Manhattan.[25]

So dominant was the General Electric Company in its industry, especially in the lamp sector, that a competing firm, the National Electric Lamp Company, was created as an independently managed subsidiary in 1901 at Cleveland. Clearly, the American lamp industry was an oligopoly. In 1910, General Electric controlled 42 percent of the American market, National Lamp, 38 percent, and Westinghouse, 13 percent.[26] Uniform prices were set—the standard 16-candlepower carbon filament lamp selling for 17 cents, one cent below its 1896 price.[27]

APPLICATIONS INDOORS AND OUT

Incandescent filament lamps improved steadily, with major advances including the adoption of carbonized cellulose filaments in 1891, metalized filaments in 1905, and metallic filaments in 1906. The big breakthrough involved the use of tungsten, specifically filaments of pressed tungsten, adopted in 1907, and drawn tungsten, adopted in 1911.[28] Tungsten, one of the heaviest elements, had been discovered in 1870. It proved 50 percent heavier than mercury and nearly twice as heavy as lead, and tungsten was thus an excellent material for electrical resistance. However, tungsten slowly evaporated when heated, blackening lamp globes and diminishing their brilliance. When inert gases such as nitrogen, argon, krypton, and helium were added, tungsten oxidized less rapidly, but luminous efficiency was reduced. Coiling the filament into a short helix and using nitrogen (and later nitrogen and argon) as gaseous fill corrected this problem. After 1914, tungsten lamps varied from 10 to 13 lumens per watt.[29] The new ductile-tungsten lamp was patented and sold by both General Electric and National Lamp under the "Mazda C" label. Ahura Mazda was the ancient Persian god of light, and the C indicated that the marketed bulb was the third prototype developed. Originally, only two sizes, 750 and 1,000 watt bulbs, were sold. Bulbs of lower wattage were then developed down to 40 watts. Other notable changes included the 1919 "tipless" bulb, with a smooth rounded top, and the 1925 frosted bulb. The life span of incandescent filament lamps was about 500 hours when first

Brightening the Lives
of your
Children's Children

WONDERFUL as is its record of triumphs, MAZDA Service strives toward even higher accomplishment in electric lighting.

For the mission of MAZDA Service is to develop not merely a better lamp, but the best illuminant that mankind can devise. For this, a corps of scientific pioneers in our Research Laboratories at Schenectady delves unceasingly into the hidden ways of science—exploring the whole world for new materials, new methods, new thoughts and supplying the results of this search to the makers of MAZDA lamps so that they may bring the perfect light always a little closer.

For this, too, thro' all the years to come, MAZDA Service will go on and on, ever seeking to improve the lamps of tomorrow as it has improved the lamps of yesterday. And thus as the mark MAZDA etched on a lamp means to you the best lamp of today, so to your children's children, MAZDA will mark the lamp that sums up in their day all this endless search for the perfect light.

GENERAL ELECTRIC COMPANY

MAZDA

"Not the name of a thing but the mark of a Service"

4616

FIGURE 3.7. Advertisement for the General Electric Company. The new "mazda" electric incandescent lamps introduced in 1914 represented the culmination of decades of experimentation focused on increasing brightness, extending bulb life, and reducing manufacturing costs. From *Sunset Magazine*, Feb. 1915, 351.

marketed. Life spans up to 3,000 hours were quickly achieved, although luminance fell to below 80 percent efficiency over time as bulbs burned out.[30] Lamps of 1,000-hour duration were made standard, a compromise between duration and quality of light.[31]

Incandescent filament lighting was developed primarily for interior use, as a light source that could simulate indoor use of gas. At first, incandescent lamps were not always available for outright purchase. In Chicago, Samuel Insull's Commonwealth Edison initially had made tungsten lamps available only to commercial customers, renting out special four-socket fixtures for shop and office use.[32] Until the development of tungsten filaments, incandescent lamps found little use in outdoor public lighting, where electric arc and gas mantle fixtures reigned supreme. Of course, there were exceptions. In 1899, Flint, Michigan, installed 650 incandescent filament lamps in a series of arches built across the city's principal retail street.[33] Other cities and towns installed the lamps in clusters on decorative posts—the so-called "white way" lighting. An early column system in St. Paul, Minnesota, consisted of five globes mounted on each column, the larger globe at top containing one 50-candlepower and five 10-candlepower lamps. Each of the four pendant globes, supported below on projecting brackets, contained four 10-candlepower lamps. Wattage per post stood at approximately 800. The top lamps, wired on a separate circuit, burned all night, while the others were turned off at midnight.[34] Eventually, high-wattage tungsten filament incandescent lamps, with their more balanced light and higher efficiency, proved competitive with arc lamps as street lights.

In 1919, San Francisco upgraded the lighting of Van Ness Avenue between Market and Valejo Streets with tungsten bulbs. The old lighting along one part of the street had been gas—three gas mantles clustered atop three posts per block, provided upwards of 500 candlepower. One arc lamp per block provided up to 300 candlepower along the remainder of the street. Tungsten lamps, each 250 candlepower, were placed two to a column, with sixteen columns per block, radiating about 4,000 candlepower.[35] A trade journal editorialized: "Under the old system of lighting it was dangerous for a pedestrian to attempt to cross the street because of the heavy automobile traffic. Now the entire street is flooded with evenly distributed light and the appearance of the street as well as the public safety has been greatly enhanced."[36] The more uniform distribution of light obtained by using large numbers of such incandescent filament lamps reduced both the depth of shadow on city streets and the amount of glare.[37] Between 1912 and 1917, the number of incandescent filament lamps used in street lighting in the United States increased from 682,000 to 1,389,000, while the number of electric arc lamps decreased from 348,600 to 256,800.[38]

AMERICANS TEND TO EQUATE the start of the electrical age with Thomas Edison's development of incandescent filament lighting. Certainly, his incandescent system revolutionized indoor domestic lighting. It was clean and easy to use, and the technology was widely affordable. It was rapidly adopted by the vast majority of America's urban households. Incandescent filament lamps were also rapidly adopted in stores, offices, and factories. Here was a light that could be "subdivided," eliminating excessive brightness and glare. But brightness was a desirable attribute in outdoor illumination, at least to the extent that glare could be abided. Until the early twentieth century, weak incandescent filament lamps could not compete with arc lamps and gas mantles in brightly illuminating major city streets. Only when clustered on columns or arrayed in strings on arches did tungsten filament lights prove competitive. Nonetheless, once tungsten lamps were first used extensively in street lighting, still other innovations soon appeared to challenge. Various discharge lamp technologies emerged and were improved by stages, ultimately displacing incandescent filament lamps outdoors. Of course, Edison's bulbs were not displaced indoors, and therefore incandescent lamps provide the light most familiar to most of us today.

The rise of electric incandescent lighting drove the electric lighting industry to maturation. Corporate entities emerged, and relationships were configured between those entities, which would predominate throughout the twentieth century. A standardized electrical lighting system for both indoor and outdoor lighting was brought to the fore and replicated locality to locality across the United States, at least in the nation's cities. The lighting of public spaces was increasingly standardized city to city. These were developments fully integral to the nation's economic growth and modernization.

CHAPTER FOUR

Cutting Costs and
Brightening the Night

The cost of lighting public streets soared with the coming of the automobile. Faster vehicles, higher traffic densities, and the more extensive mixing of vehicles and pedestrians made higher levels of illumination necessary, especially in city business districts and along major commercial thoroughfares. No longer adequate was the pooling of light at intervals along streets. A strictly even illumination was now needed, in which vehicles and pedestrians could be fully lit against backgrounds of street surface and building facade. In addition, minimal levels of illumination had to be maintained throughout the night. No longer could city lighting systems function only on moonless nights or only until midnight or some other designated hour.

Lighting was also an activity closely tied to policing. That its benefits for nighttime surveillance had reduced crime against both persons and property was a belief little challenged. Thus municipalities used public lighting as a means of extending the effectiveness of their police forces, light substituting for more costly increases in manpower. By 1910, lighting had come to represent a substantial capital outlay for municipal governments. Ways of reducing costs had to be found.

Street lighting systems were expensive to install, and, more importantly, they were expensive to operate and maintain. Many cities sought to lower lighting costs through municipal ownership, Chicago being the largest city to operate its own system.[1] At first, municipal ownership did lower costs. Detroit's system, championed by reform mayor Hazen Pingree in the mid-

1890s, lowered lighting expenses by nearly half.[2] Very quickly, however, the economics of power generation and distribution shifted in favor of regulated private utility companies. Many cities, nonetheless, maintained ownership of street light fixtures even after they gave up power generation. A 1954 survey of 276 cities in the United States revealed that seventy-two still owned their own public lighting fixtures and forty-nine still owned at least a portion of them.[3] With the rise of large central power stations and with the introduction of differential rate structures, however, commercial power became much cheaper than publically generated electricity. Street lighting became an important means by which private utility companies equalized load curves across diurnal cycles.

Improvements in power generation, largely through increased scale of operation, were matched by improvements in lamp efficiency. General Electric, Westinghouse, and several other companies established research programs dedicated to developing cheaper alternatives to arc and incandescent filament lamps. Out of this activity came the electric gaseous discharge lamp—a category of light fixture that ultimately included mercury vapor, sodium vapor, neon, fluorescent, metal halide, and even laser lights. By 1950, the new lamps had almost totally displaced their closely related prototype, arc lighting. After 1950, discharge lamps also began to supersede incandescent lamps in street lighting as the replacement bulb of choice.

The first experiments with glowing gases were conducted in Germany by artist and glassblower Heinrich Geissler. When he passed a high-voltage alternating current through a sealed tube containing air at low pressure, the tube gave off a weak luminosity until its vacuum failed.[4] Soon after, the light-rendering properties of a wide array of gases were tested by researchers in a number of countries. Around 1890, Thomas Edison experimented with calcium tungstate, from which, under an electrical charge, he obtained fluorescence. The idea behind the discharge bulb was quite simple. When gas pressure inside a sealed tube was low, an electric current could be passed through the tube between electrodes located at each end. Voltage applied at the electrodes accelerated the flow of free electrons, which bombarded the atoms of the gas, displacing electrons from their normal atomic positions. As the displaced electrons fell back to their normal positions, energy was radiated.[5] Radiation varied from the ultraviolet to the visible to the infrared as a function of the gas used and the degree of its displacement.

MERCURY VAPOR LIGHTING

The idea behind the mercury vapor lamp originated in the work of D. McFarland Moore, one of Thomas Edison's assistants from the Menlo Park era. Conducting his own research with Geissler tubes, Moore sought to improve upon Edison's experiments. In 1899, he equipped New York City's Madison Square Garden with a 186-foot glass tube two inches in diameter, which bathed the Garden's foyer with a bright white light.[6] Moore's lamp tubes required extremely high voltages, however, and they were very complicated to install. They did not represent a standard fixture for mass consumption so much as a custom installation for special use. Although proposed for street lighting, Moore's tubes never enjoyed such application (fig. 4.1). Nonetheless, they invited continued experimentation with outdoor use in mind. In 1912, Moore sold his patents to General Electric and assumed a research role at the company's Schenectady laboratory.[7]

At General Electric, researchers discovered that electric current passed in only one direction in mercury vapor, and they set to work perfecting a mercury-arc rectifier that converted alternating current into direct current. Charles P. Steinmetz, whose personae largely replaced that of Edison's as General Electric's scientific genius, applied these discoveries in creating a tubular vapor lamp. At the same time, Peter Cooper-Hewitt, with funding from George Westinghouse, perfected a four-foot-long tube lamp in which an electric arc in mercury vapor produced an eerie greenish-blue light, at about 12.5 lumens per watt.[8] As the U.S. Patent Office was unable to establish clear priority of invention, General Electric and the newly formed Cooper-Hewitt Electric Company agreed in 1913 to exchange patent licenses, General Electric buying out Cooper-Hewitt six years later.

The early Cooper-Hewitt lamp was inclined at a ten to fifteen degree angle when operated, in order to keep the arc from burning through its glass surround (fig. 4.2). Both ends of the glass tube were enlarged, and the lower end contained a reservoir of mercury which acted as a cathode. The upper end served as the anode and allowed for vapor expansion. The

CAL ENGINEER.   [Vol. XXIV. No. 485.

MOORE VACUUM TUBES FOR STREET AND CAR LIGHTING.

IN the accounts we have given up to the present of Mr. D McFarlan Moore's system of vacuum tube lighting, the tubes were employed for interior illumination. Looking toward the future, however, Mr. Moore has recently tried som experiments in outdoor vacuum tube illumination, which lead him to the belief that such work is in every way practicable

STREET LIGHTING BY VACUUM TUBES FROM RAILWAY CIRCUITS.

In a recent out-of-door experiment of this kind, it was foun that with the light of only one tube the seconds' hand of watch could be read at a distance of 60 feet. The tube i _____ was similar to those Mr. Moore has used in previou

FIGURE 4.1. Proposal for use of Moore vacuum tubes for street lighting. Although not adopted, the anticipated visual effect of Moore's lighting was not unlike that eventually achieved with fluorescent tubes. From *Electrical Engineer* 24 (1904).

lamp had to be tilted by hand to start: a flow of electric current heated mercury released from a reservoir, vaporizing some of it to create an arc. Hand operation was eliminated through use of electromagnetic tilting mechanisms, heating coils, and sparking devices, but the Hewitt lamp remained a complicated device, awkward to use. Also, the light produced was almost totally lacking in red wavelengths, rendering people and things "unnatural" in appearance. Initially, therefore, mercury vapor lamps almost always were used in combination with incandescent lamps set to radiate excessive redness as a balance to the mercury vapor lamps' greenish-blue.[9] A standardized 400-watt low-pressure bulb, radiating some 16,000 lumens of light, was developed for street lighting.[10]

High-pressure mercury vapor lamps were developed in the 1930s in Europe by three companies working independently—Philips in the Nether-

D. C. 55 Volt 3.5 Amp. Two in Series on
100-125 Volt Circuit

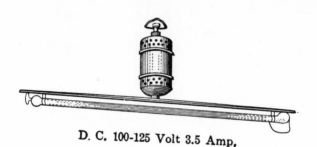

D. C. 100-125 Volt 3.5 Amp.

A. C. 100-125 Volt 4.1 Amp.

FIGURE 4.2. Variations of the Cooper-Hewitt gaseous discharge lamp. Light was produced by electric arcing in gases within sealed tubing. These fixtures required hand tilting to ignite, limiting their usefulness for outdoor illumination. From *Illuminating Engineering Society, Illuminating Engineering Practice* (New York: McGraw-Hill, 1917), 158.

lands, Osram in Germany, and General Electric's British subsidiary.[11] These lamps incorporated tungsten wire electrodes in special glass tubes about thirteen inches long and two inches in diameter. As with low pressure lamps, they were hung vertically; when placed horizontally, the long

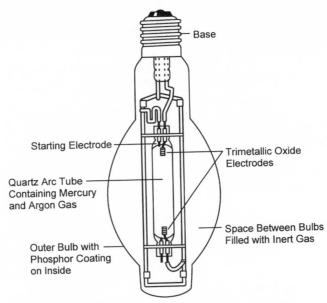

FIGURE 4.3. Basic elements of a modern high-pressure mercury vapor lamp. Sealed under vacuum with an electrical charge arcing between electrodes, gaseous discharge lamps were related to the earlier electric arc lamps. Arcing involved various gases calculated to intensify brightness. Modified from *Roadway Lighting Handbook* (Washington, D.C.: U.S. Department of Transportation, 1978), 40.

arcs tended to bow and touch the glass tubing, melting it. Magnets designed to draw the arcing away from tubular surrounds eventually enabled horizontal placement. In some early lamps, tubes containing the arc were surrounded by outer glass globes through which water circulated to maintain internal temperature and air pressure. The fixtures were cumbersome in appearance and difficult to install owing to their bulkiness. In the 1950s, after durable, fire-resistant synthetic quartz had been developed, the magnets and their water surrounds were eliminated, making the lamps suitable for streetlight fixtures.

Mercury vapor lamps found widespread use in the United States only after 1950. A critical factor was the lengthening of lamp life to roughly 16,000 hours. Since about half the energy radiated by the mercury arc was in the ultraviolet region, phosphor coatings on the inside of the outer bulb were adopted to produce red light, giving a more balanced luminance without use of a parallel incandescent filament bulb.[12] Yet, as phosphor coatings broke down with use, lamps lost much of their color value over time. As burnt out bulbs were replaced in a lamp array, a range of different color tones could develop, lamp to lamp. Mercury vapor lamps also contained argon to aid in ignition; a starting electrode heated the argon to produce a muted blue glow. Heat from this small discharge vaporized some of the mercury, which increased internal pressure and caused the arc to strike. Some four to seven minutes were required for mercury lamps to reach optimum light output, an effect that added visual interest to the twilight.[13]

To early advocates, light from mercury vapor lamps had a remarkable "revealing power," as if the peculiar light spectrum tended to amplify visual acuity. "The eye can see with less light of the simple composition of mercury vapor light than it can with the complex light of ordinary sources," wrote E. L. Elliott in the *Electrical Review* in 1921. "This peculiarity, reduced to practice, means that there are no black impenetrable shadows under properly installed mercury vapor lighting." For lighting engineers, mercury vapor promised, Elliott asserted, a virtual "obliteration of troublesome shadows."[14] As the new lamp produced no red rays, it was seen to be, in its unmodified form, more restful on the eyes. Motorists could see with less effort, and fatigue was relieved. Reaction time was reduced. On the down side, mercury lighting produced unsatisfactory glare. Refractors for eliminating the harsh light were difficult to affix, as the lamps already had a bulky outer glass surround. The main objection involved the sallow greenish light, which turned nighttime streets pale. Whereas green and blue objects were amplified in hue, red or orange objects appeared dull brown.

Ultimately, it was the cost efficiency of mercury vapor that brought the technology to the forefront in American street lighting. By 1950, mercury vapor lamps could deliver forty lumens per watt, compared with sixteen to twenty-one for incandescent lamps. The first sizable street lighting installation in the United States was made on Denver's Park Avenue in 1936, where the vapor bulbs were installed along with incandescent lamps.[15] Applications increased slowly; only 250,000 mercury vapor street lamps were in service nationwide by 1954.[16] Ten years later nearly 39 percent of all streetlights were of mercury vapor, incandescent and fluorescent lamps accounting for 60 percent and 1 percent, respectively.[17] Americans quickly adjusted to the new light, which, although harsh, made incandescent lighting appear dim in contrast. Mercury vapor fit the American penchant for the new, as well as the perception that brighter was inherently better.

SODIUM VAPOR LIGHTING

Until the early 1930s, there was no type of glass that was fully resistant to the erosive qualities of sodium vapor. In 1931 such a glass was developed in Germany, and in 1932 a sodium vapor street lamp was introduced in the Netherlands.[18] Initially, sodium vapor lamps consisted of sixteen-inch tubes about three inches in diameter, with coiled oxide-coated filaments at both ends, each serving as cathode, while a sleeve of molybdenum at one end served as an anode. Tubes contained sodium and neon gas, the latter used as a starter. An outer, double-walled glass surrounded tubes to control temperature and provide pressure seal. Early lamps were very slow starting, requiring up to thirty minutes to come to full yellow luminance. In the interim, the neon provided a reddish glow. A 220-watt lamp could produce 10,000 lumens—equal to the output of a 550 watt incandescent bulb.[19]

Sodium vapor produced a strong, nearly monochromatic yellow light,[20] which, being near the maximum sensitivity level of the human eye, proved useful in revealing the details of objects at low levels of luminance. Sodium vapor lamps, therefore, enjoyed early application in factories and workshops. But the lamps appeared to change the color of objects to either yellow or black. Like mercury vapor lamps, they gave people and things a pale, washed-out appearance. The mixing of gases in high-pressure sodium fixtures, however, enabled some color balance, and most modern sodium vapor lamps produce a golden-pink light.

In the United States, sodium vapor street lights were first installed in 1933 near Port Jervis, New York, along a rural highway.[21] Through the 1930s, lamp efficiency was not sufficient to make sodium vapor competitive with incandescent lighting, which, of course, had the further advantage of appearing more natural. Enhanced visual acuity under sodium vapor light did make it recommendable for safety lighting, especially on bridges, in tunnels, and at the "cloverleaf" interchanges of new limited-access express highways. Yellow was thought to encourage caution, and cities began to experiment with the new light accordingly. Chicago lit selected intersections in outlying business districts, and the new lighting was touted as reducing the number of accidents in those areas by about 17 percent over the first five months and reducing personal-injury accidents by about 46 percent.[22] Such statistics, however, rarely reflected careful comparative analysis.

Sodium vapor light was also valued as a crime-fighter. The distinctive yellow glow was promoted by the lamp manufacturers as a warning sign to perpetrators in high-crime areas. But cities like Newark and New Orleans rejected sodium vapor lights *because* of this association. Few cities would accept the stigma of acknowledging high-crime neighborhoods. After its promising introduction, sodium vapor rapidly declined and mercury vapor remained the preferred gaseous discharge lamp for outdoor lighting. Sodium vapor lamps were not even available commercially in the United States for a period in the early 1960s.[23] But the introduction of high-pressure sodium lamps, promoted once again for the crime-fighting implications of their yellow light, reversed this trend. The riots of the late 1960s brought great attention to the issue of street surveillance. In New York City, in 1975 alone, 1,200 miles of major thoroughfare were "relit."[24]

Nonetheless, lower costs, and not crime-fighting potential, proved the most important incentive in adopting sodium vapor lighting. In 1980, incandescent filament lamps cost, on average, $280 per lamp per year to operate as streetlights. A mercury vapor lamp's average annual cost was calculated at $128. But the yearly operating costs of low-pressure and high-pressure sodium vapor lamps were calculated at $60 and $44, respectively.[25] Cost differentials reflected not only lamp efficiencies but also increased life expectancies. With standard lamp life set at 15,000 hours, high-pressure sodium vapor required substantially less maintenance with substantially lower labor cost. Consequently, municipalities could afford to ignore the bland and drab visual scenes that the lamps produced. Americans would have to adjust themselves to nighttime landscapes tinted golden pink.

NEON LIGHTS

Heinrich Geissler's tube lamp of 1856 had one major drawback. Energized gas reacted chemically with the electrodes, which rapidly deteriorated and resulted in a loss of pressure and subsequent lamp failure. D. McFarland Moore's tube lamp, developed in 1904, solved the electrode and pressure problem, but it was Frenchman George Claude who carried McFarland's experiments forward using distillations of rare gases such as neon and argon. Neon, only discovered in 1898, was an inert, colorless gas; but when excited electrically, it glowed a bright orange-red. Argon was found to glow a bright grayish-blue. Further experiments followed, yielding a virtual rainbow of colors. Blue resulted when mercury was added to argon. Green resulted when the same mixture was placed behind yellow rather than clear glass. White was created using helium, and yellow using helium under yellow glass.[26] The word *neon* became a generic term descriptive of the entire spectrum of glowing tubes produced.

George Claude offered General Electric a license to manufacture and market his tubes in the United States, the devices having received extensive use in Europe in advertising signs. But General Electric's directors, seeking standardized products, miscalculated neon's potential and thought that incandescent filament bulbs would continue, as in the past, to dominate outdoor advertising. Consequently, Claude established his own company, Claude Neon Lights, Inc., to license small sign companies across the nation, companies that specialized in custom sign installation. The art of bending glass tubes into all manner of shapes for lettering and various ornamental schemes quickly developed. America's first neon sign—the word *Packard* spelled out in orange—was installed in 1923 at a Los Angeles automobile showroom.[27] Yet it was not until Claude's principal patents expired in 1932 that neon signs truly became prominent in the American night, especially in big city business districts and along small town main streets.

FLUORESCENT LIGHTING

Fluorescent lamps are also a kind of arc lamp, with electrodes placed at the ends of a glass tube between which current jumps to excite mixtures

FIGURE 4.4. Neon sign near downtown Louisville, Kentucky, 1997. Neon is used almost exclusively in outdoor advertising. It attracts the eye, giving streets points of colorful visual focus.

of mercury and argon gas. Some of the energy radiated by fluorescent lighting is in the blue spectrum, but most is in the ultraviolet range. Interiors of the glass tubes are therefore coated with powdered phosphors that glow, or fluoresce, under ultraviolet radiation to produce visible light. Some principles of fluorescence were known in the seventeenth century, although it was not until 1852 that Englishman George Stokes discovered ultraviolet light induced fluorescence in various substances. Initially, he found that quinine sulphate absorbed invisible ultraviolet light to give off visible radiation at larger wavelengths.[28] He named the effect after the mineral fluorite, which he found exhibited the strongest tendency to give off visible light. The first lamp containing fluorescent material was developed a few years later and suffered, as many early lamp innovations, from low efficiency and very short life-span.

Thomas Edison's 1896 experiments with tube lamps led to General Electric's research with fluorescing phosphors. But it was American inventor Gilbert Schmidling, working independently in 1934, who actually perfected the use of phosphor-coated glass bulbs lined with fluorescent

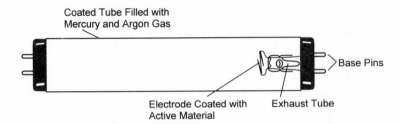

FIGURE 4.5. Basic elements of a fluorescent lamp. An electrical charge excites otherwise inert gases in coated glass tubes, which are manufactured in various standard lengths and in various shapes, including circular lamps.

powder. Fluorescence was produced by electron emission from an oxide-coated sphere at the center of the bulb.[29] Researchers at General Electric, meanwhile, perfected a safe electric-discharge device for tube use which led to the first commercially feasible fluorescent tubes in 1935. A line of lamps varying from 18 to 30 inches in length was placed on the market in 1938.[30] The "new light" came in various hues, reflecting the various phosphors or phosphor mixes used. Sales soared from 200,000 bulbs in 1938 to 21 million in 1941.[31] Highly efficient at fifty lumens per watt, standardized 48-inch lamps found ready use for indoor lighting, especially in retail stores, offices, hospitals, factories, and wherever a flood of brilliant, even light was desired.

Fluorescent lamps, although applied early to outdoor floodlighting, were slow to be adopted for street lighting. Fluorescent light radiated from long tubular surfaces, and, accordingly, could not be easily focused. Luminaires were bulky and not easily handled, inflating labor costs. Their weight proved excessive for older lampposts, and so they could not serve readily as replacement lamps. Only in Europe, where high energy costs coupled with low labor costs created an economic imperative, did the technology see immediate adoption for street lighting. The first municipally financed installation of fluorescent streetlights in the United States occurred in 1950 along Detroit's Wyoming Avenue, an evolving, and previously poorly lit, commercial strip on that city's far west side.[32] In 1954, Chicago's State Street merchants substituted fluorescent lamps for incandescent lights, making State Street some four times brighter.[33] By 1970, many cities had adopted fluorescent lights as a means of upgrading declining retail districts.

Fluorescent street lighting, like mercury and sodium vapor lighting, cast a light that most Americans, conditioned by incandescent filament lamps, viewed as too artificial. Nonetheless, fluorescent lighting's differences from daylight were not as exaggerated as those of other gaseous discharge technologies. Colors were less distorted, objects appeared less angular, and human features were less depersonalized. The greatest drawback to using fluorescent lamps remained their size, which necessitated new lampposts. Therefore the adoption of fluorescent lamps required the total rebuilding of street lighting infrastructures. Fluorescent lighting did, however, find ready application where outdoor systems were built from scratch, as, for example, in the parking lots of new suburban shopping centers.

METAL HALIDE LIGHTING

The rapid return to sodium vapor lighting in the 1970s reflected, in part, that technology's relatively softer light in comparison to mercury vapor. Efforts to modify the mercury vapor light then produced the metal halide lamp. This lamp, first called the multivapor lamp, began as a mercury vapor prototype altered to use various iodine compounds.[34] It gave off an exceptionally brilliant white light closer to daylight at efficiencies near ninety lumens per watt, and life expectancies were up to 10,500 hours.[35] Metal halide lamps required special ballasts capable of exceptionally high voltages, and they were equipped with more substantial electrodes than were previously used. But overall, metal halide fixtures came to look very much like mercury vapor lamps.

ELECTRIC GASEOUS DISCHARGE LAMPS came to displace electric arc and then electric incandescent filament lamps in outdoor lighting, at first adding to the diversity of night light in cities and then overwhelming that diversity with degrees of homogeneity. Mercury vapor and sodium vapor, and even fluorescent lamps, found widespread application in street lighting thanks to their cost-efficiency and brightness. Their luminance now dominates the streets of American cities, turning them greenish-blue or golden-pink in arrays of linear brilliance. Gaseous discharge technologies built upon earlier discoveries and, like all other electric lighting, did not suddenly appear "out of the dark." Applications reflected not just felt needs but also, more importantly, economic realities. Adoption, however, was always modified by popular tastes established by previous lighting systems. Thus mercury and sodium vapor lamps were at first thought appropriate only for locations where color distortion was of little consequence (along rural highways, on bridges, or in tunnels), or where, generally, there were few pedestrians. But, as Americans increasingly took to nighttime motoring and their experience with such light increased, the distortions of gaseous discharge technologies became commonplace.

Among the new gaseous discharge lamps developed, neon stood out and was applied to outdoor advertising rather than to the lighting of city streets. Neon's popularity for sign art grew rapidly through the 1930s, falling off after World War II. Installation and maintenance costs were high, and, visually, the dazzle of neon signs against backgrounds of shadow was much reduced amid the surrounding brilliance of mercury vapor and sodium vapor street lights. Automobile use, and the intensive street lighting it required, would dominate the illumination of cities at night. Although fluorescent lighting was mainly applied to indoor use, fluorescent tubes in "backlit" signs would prove to be neon's principal replacement, being both cheaper and easier for rapidly moving motorists to read.

PART II

LIGHTING THE AMERICAN CITY

Application and
Change

CHAPTER FIVE

Lighting City Streets

As lamp technologies evolved, so also did expertise in light's technical applications. Street lamps—whether oil, gas, or electric—were set in increasingly sophisticated fixtures, or luminaires, to amplify and direct the light as appropriate to changing needs. With the coming of electric arc lamps, especially, a new kind of expert—the illuminating or lighting engineer—evolved to perfect lighting procedures and set lighting standards. What began as a imprecise art of trial and error eventually grew into a science of lighting practice with carefully articulated engineering standards. The automobile was central to this new science. The most pressing needs in traffic engineering were to make the nighttime safe for motoring and to make the traditional street, as changed by the presence of the automobile, safe for pedestrians. How much light was needed? How should it be projected? How should lamps be arranged? As cars came to be fitted with their own lighting, how should these lamps be regulated? Street lighting continued to serve other purposes, of course, but its role in facilitating vehicular movement would remain primary.

Standards promoted by design professionals changed by stages, as questions were repeatedly asked and answered in the context of new lamp technologies and changing economic and social contexts. Although rarely asked as such, the important question should have been, How should American cities be seen and experienced in the night? Instead, the lighting engineers asked, What were the minimum light levels necessary for the performance of given tasks? Direct sunlight varies from 5,000 to 10,000 foot-candles in strength, while natural nighttime levels of light vary from approximately 0.0002 foot-candles for starlight to 0.02 foot-candles for moonlight.[1] Recommended street-lighting levels today vary from 0.2 to 2.0 foot-candles. In contrast, recommended lighting for offices varies from 100 to 300 foot-candles. Although the human eye can function across this entire range of illumination, specific tasks do require certain minimal light levels. When available light falls below 0.003 foot-candles, cone, or foveal, vision is no longer useful and the human eye cannot perceive color or fine detail.[2]

LIGHTING STANDARDS

Street-lighting parameters were first established in 1914 by the National Electric Light Association and the Association of Edison Illuminating Companies. Specifically, lighting engineers were charged with rendering large objects and surface irregularities in streets and on sidewalks discernible.[3] To accomplish this, deep shadows needed to be eliminated and distracting glare needed to be reduced. Illumination had to be evenly distributed along streets, enabling motorists, most especially, to keep attention levels high. Streetlights themselves were not to compete for attention but rather remain primarily as background. The purpose of lighting was not only to illuminate but also to orient. It had to cue easy comprehension of the otherwise darkened environment, and it had to do so unobtrusively.

Beyond traffic management and the subsidiary concerns of public safety and business promotion, lighting also was intended to enhance overall nighttime ambiance, contributing to a strengthened sense of place. Lamp manufacturers and power companies constantly emphasized in advertising the place-creating aspects of street lighting. "I make your streets brighter, your town prosperous," declared one advertisement placed by General Electric in the *American City,* the trade journal for city administrators (fig. 5.1). Street lighting could be a kind of "show window" for a city.[4] It could advance city pride, demonstrate competent city administration, and foster other civic improvements. Lighting could drive up property values and attract investment from the private sector.[5] In various ways, street lighting was made to stand for progress. Bright light at night, General Electric's ad advised, made for busier and more attractive cities, altogether "happier" places in which to work and reside.

FIGURE 5.1. Advertisement promoting street lighting. Companies like General Electric emphasized that light had place-making capabilities. Downtown streets, when brightly lit, seemed quintessentially up-to-date. Functionality combined with novelty to be hailed as progress. From *American City* 16 (1916): 70.

The principal forum for setting street-lighting standards became the Illuminating Engineering Society, formed in 1906 to advance "the theory and practice of illumination" and disseminate "knowledge relating thereto."[6] The Society's first president was Louis B. Marks, inventor of the enclosed carbon arc lamp. Indeed, Marks was, perhaps, the first person to identify himself in the United States as an illuminating engineer.[7] In 1925, the society's Highway Lighting Committee was formed to establish the scientific principals underlying street and highway lighting, to collect data on the results of the application of such principals to actual practice, and to disseminate findings in reports "readily accessible to the public."[8] Appointed to the committee were persons representing the lamp manufacturers, the public utility companies, government at the federal, state, and municipal levels, the growing number of consulting engineers, and university educators.

Street-lighting standards reflected the following physical qualities: (1) the intensity of the light source and its brilliancy (or the flux density at the surface of the illuminant), (2) the flux of light (or the total visible radiation produced by the illuminant), as measured in lumens, (3) the light density (or the distribution of the light flux in space), as measured in lumens per square foot, (4) the illumination (or the light flux density issuing from the illuminated object), also measured in lumens per square foot, and (5) the intensity of the light source as measured in candles.[9] Because light intensity, or brilliancy, expressed in lumens per square foot yielded enormous numbers, figures were generally expressed in lumens per square centimeter. These measurements were used to gauge the light produced or reflected in the visual environment; they did not evaluate the light as actually perceived by users in a specific location.

Illumination available in a visual environment is not the same as the light actually used by persons performing visual tasks. Recognizing the importance of "effective light," the International Commission on Illumination (CIE) instituted in the 1980s another set of street-lighting standards based on roadway luminance as opposed to illuminance.[10] *Illuminance* refers to the light falling on street pavements and other objects (what photographers call "incident light"), whereas *luminance* refers to light actually reflected to the eye. As with illuminance, luminance is affected by the following factors, which underpin street-lighting practice:

(1) lamp type, (2) luminaire design (including size), (3) mounting height, (4) fixture spacing or positioning, (5) street width and surface material, (6) weather conditions, (7) traffic type and density, and (8) the built environment lateral to a street, including competing and complementary sources of illumination.

LUMINAIRES

In a broad sense, the term *luminaire* can refer to both the street lamp as light source and to its support and enclosure. My focus is on the lamp surround. In this narrower use of the word, the luminaire directs the light as it supports and protects the lamp, connecting the lamp to its energy source. Luminaire design and positioning represented the very essence of streetlight application. Early on, luminaires merely held candles or oil lamps in place, and lantern glass was added to provide protection for the flame. Luminaires later connected gas lamps to gas pipes and included lantern surrounds, often with reflecting mirrors to direct the light. In electric incandescent filament lamps, luminaires included glass surrounds constructed to refract light, further concentrating it and enhancing its effectiveness. In the first electric arc, as well as the earliest electric gaseous discharge lamps, luminaires were not enclosed but left open. Later, after electrodes and ballasts had been much reduced in size, reflectors and glass refractors were added.

First and foremost, luminaires control the distribution of light.[11] Light control is achieved through reflection, refraction, and diffusion, or a combination of two or all three (figs. 5.2 and 5.3). With reflectors, lost or poorly utilized light is captured and redirected to the street. Specular reflectors are made from glossy materials, such as polished porcelainized steel and aluminum, which provide mirrorlike surfaces to focus light. Diffuse reflectors, on the other hand, spread light over large areas rather than focusing it into tightly controlled patterns; plastic and painted aluminum are commonly used today for diffuse reflectors. Refractors have traditionally been made of glass, and boro-silicate glass compounds or acrylic and polycarbonate plastics predominate today. A refractor is of prismatic construction and bends the light passing through it, focusing the light outward and downward into desired illumination patterns.[12] Light is not

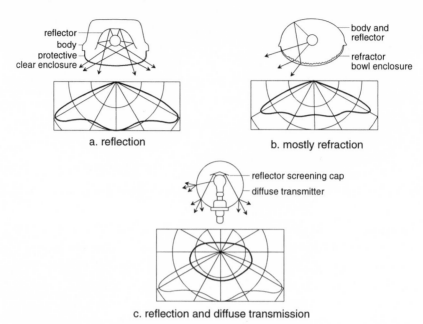

a. reflection

b. mostly refraction

c. reflection and diffuse transmission

FIGURE 5.2. Reflection (A), refraction (B), and reflection and diffuse transmission (C). In streetlights, luminaires are variously equipped to direct light. Shown here are three luminaire configurations and the associated light patterns cast in vertical cross-section. Modified from I. W. J. M. Van Bommel and J. B. de Boer, *Road Lighting* (Antwerp: Philips Technical Library, 1980), 98.

merely emitted by lamps, it is also variously focused or diffused through luminaires, and the manner by which this is accomplished impacts the look of cities at night.

Vertical and lateral light distributions were systematically classified, and modern street light luminaires are manufactured to meet the specifications of one or another classification (fig. 5.4). Arc and incandescent filament lamps employed asymmetrical prismatic devices with reflecting prisms on one side and combinations of diffusing ribbings and redirecting prisms on the other.[13] Light flux that otherwise diffused upward or onto adjacent building facades could be directed downward onto street surfaces. The first systematic use of such refracting and diffusing globe devices occurred in Milwaukee in 1916.[14] Luminaires there were also

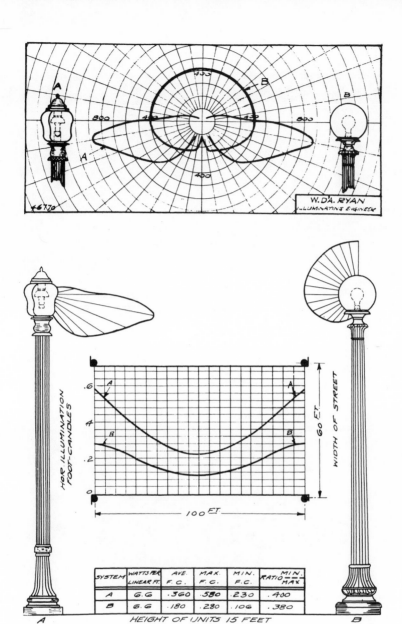

FIGURE 5.3. Lamps with and without light-deflecting globes. Shown above are the patterns of light thrown by each lamp. Shown below is the related light intensity measured in foot-candles. On the left luminance is substantially deflected down on the street; on the right it is not. From S. L. E. Rose and H. E. Butler, "Single Light Compared with Cluster Units for Street Lighting," *General Electric Review* 22 (Dec. 1919): 1044.

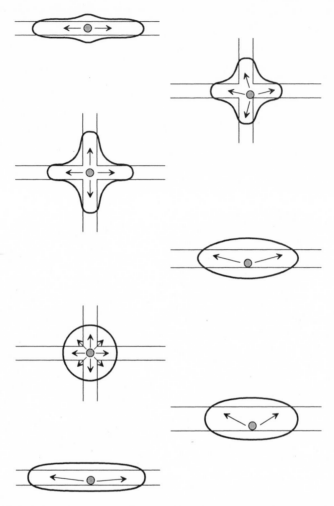

FIGURE 5.4. Selected categories of approved roadway coverage. Luminaries bend light in various spatial patterns to amplify street illumination. From *Roadway Lighting Handbook* (Washington, D.C.: U.S. Department of Transportation, 1978), 54.

mounted on thirty-foot posts, which were much taller than those previously used, making that city's installation the forerunner of modern street lighting on several counts. Glare from street lighting was controlled by mounting luminaires well above motorist and pedestrian lines of sight. Additional control was gained by means of reflectors that shielded and redirected the light to angles below 75 degrees from the vertical.[15]

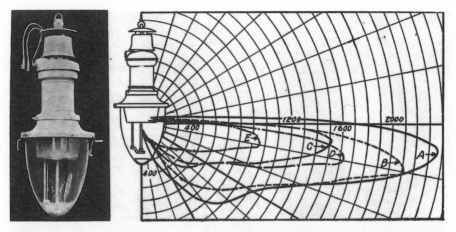

FIGURE 5.5. A pendant-style luminous arc lamp with prismatic refractor. Luminaire design and positioning represented the very essence of street light application. From S. L. E. Rose and H.E. Butler, "Modern Electric Street Lighting," *American City* 18 (March 1918): 244.

The glass surround was a luminaire's most noticeable feature, night or day. Apart from the light it emitted, a luminaire's size, shape, surface texture, and the manner by which it was attached and supported determined its visual character. Luminaires could be placed squarely on top of posts or columns, set upright on arms extending laterally from posts or poles, or hung from such extensions, or even hung from cables suspended across streets. Primitive oil and gas lamps were usually placed in boxlike lantern surrounds directly atop posts, while open arc lamps were usually suspended from armatures or cables. Enclosed arc lamps were likewise suspended, but they could also be affixed atop posts, their acorn-shaped silhouettes introducing a design prototype also widely used with incandescent filament lamps (fig. 5.5).

For incandescent lamps, the ball globe atop a post, or its closely related cousin, the vase-shaped globe, were popular through the 1920s (see fig. 5.3). The clustering of such lamps on columns with arms, upon which lamps either hung as pendants or sat upright, became the visual hallmark for numerous "white ways" established across the United States. The use of frosted glass globes without reflectors and refractors produced substantial glare; but it was that glare that gave a sense of brilliance, bringing visual excitement to a night scene. Today, many small towns and cities across the United States are reintroducing "white way" lighting as a means of enlivening "main street" business districts at night.

FIGURE 5.6. Recently installed "white way" lighting in an Illinois small town, 1998. Clustered globes light a public square, complementing a full moon above. "White way" lighting enjoyed widespread revival in the 1990s and was introduced as a means of reestablishing visual excitement after dark, especially in small towns and suburban shopping districts.

Down-turned "helmet" fixtures emerged with electric gaseous discharge lamps, which, when streamlined and combined with a curving pole, gave the impression of a serpent's head. The earliest mercury vapor and sodium vapor lamps were difficult to fit with spherical globes, and, accordingly, their irregular tubular mechanisms were left fully exposed. As mercury vapor, sodium vapor, and, later, metal halide technologies were reduced in size, more traditional glass surrounds were developed for gaseous discharge lamps. Only then were the new lamps fully appropriate as replacements for the less efficient incandescent filament lamps in their old fixtures. Fluorescent lamps, however, remained an exception. Fluorescent lamps were contained from the start in large glass-bottomed boxes appropriate to the linear tubes used. Sometimes the tubes were clustered vertically in glass tubular surrounds set at an angle on lampposts, their light deflected downward.

THE GEOGRAPHY OF STREET LIGHTING

Planners and lighting professionals recognized that different urban areas and different kinds of city streets had varying lighting needs. They also recognized that these needs varied between different sized cities. Manufacturers began to codify lighting contexts, out of which came hierarchically structured lighting designs remarkably similar across the country. In 1917, Preston Millar, in his lengthy review of then current lighting practice, offered a simple street classification scheme: (1) important and heavy traffic streets, (2) secondary business streets, (3) city residence streets, (4) suburban highways, and (5) suburban residence streets.[16] The average horizontal illumination levels recommended varied from 0.5 to 1.0 lumens per square foot for the thoroughfares with heaviest traffic to 0.005 to 0.02 lumens per square foot for residential streets in suburban areas.[17] On business streets, building facades needed to be lit as well as street surfaces. On residential streets, light needed to be restricted as much as possible to street surfaces so as not to flood darkened bedrooms.

In this way, lighting engineers began to conceptualize the American city according to street hierarchies, each street type with its own lighting solution. Lamp size, mounting height, spacing, and positioning all varied systematically from one category of street to another. Different lighting designs were proposed for cities of different sizes (figs. 5.9 and 5.10). As "rule of thumb" recommendations, such design codes took width of street and adjacent land uses as indirect measures for traffic load. By the 1940s, however, direct traffic counts were being integrated into hierarchical design schemes. In 1944, for example, Westinghouse engineers recommended that luminaire brightness, mounting height, spacing, and positioning reflect categories of actual traffic flow (fig. 5.11). Today, roadway categories are carefully defined, with lighting recommendations for each category based on the specifics of illumination depreciation outward from luminaires, as well as luminaire mounting height, spacing, and positioning.[18]

Because municipalities tended to buy lighting systems as packages, one or another hierarchical lighting scheme was adopted in most places. In 1911, Champaign, Illinois, installed a "complete ornamental system" employing "tungsten lamps."[19] Lamps of 200 candlepower were installed on

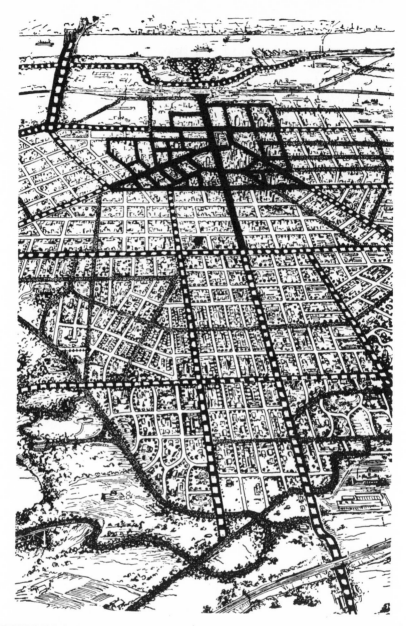

FIGURE 5.7. Lighting zones by category of street in a hypothetical city. The diagram clearly illustrates how lighting engineers categorized city streets hierarchically, with busiest thoroughfares and downtown areas lit at highest intensities. From Ward Harrison, O. F. Haas, and Kirk M. Reid, *Street Lighting Practice* (New York: McGraw-Hill, 1930), 122. Reproduced with permission of The McGraw-Hill Companies.

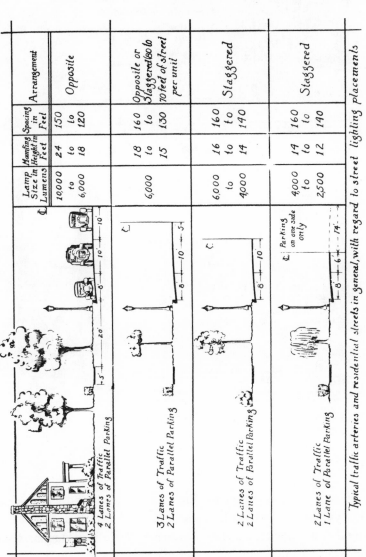

	Lamp Size in Lumens	Mounting Height in Feet	Spacing in Feet	Arrangement
4 Lanes of Traffic 2 Lanes of Parallel Parking	10,000 to 6,000	24 to 18	150 to 120	Opposite
3 Lanes of Traffic 2 Lanes of Parallel Parking	6,000	18 to 15	160 to 130	Opposite or Staggered 180 to 70 feet of street per unit
2 Lanes of Traffic 2 Lanes of Parallel Parking	6,000 to 4,000	16 to 14	160 to 140	Staggered
2 Lanes of Traffic 1 Lane of Parallel Parking	4,000 to 2,500	14 to 12	160 to 140	Staggered

Typical traffic arteries and residential streets in general, with regard to street lighting placements

FIGURE 5.8. Recommended lighting for categories of residential street. Proposed here is a hierarchy of lit streets with lamps varying from 2,500 to 10,000 lumens and with luminaires positioned at various heights and spaced at various intervals. From Charles J. Stahl, *Street Lighting* (New York: John Wiley and Sons, 1929), 19.

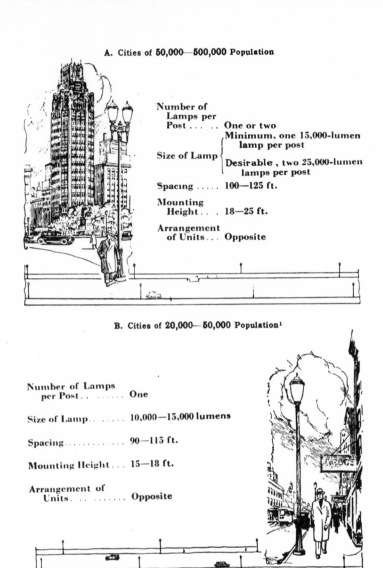

A. Cities of 50,000—500,000 Population

Number of Lamps per Post	One or two
Size of Lamp	Minimum, one 15,000-lumen lamp per post
	Desirable , two 25,000-lumen lamps per post
Spacing	100—125 ft.
Mounting Height . . .	18—25 ft.
Arrangement of Units . . .	Opposite

B. Cities of 20,000— 50,000 Population[1]

Number of Lamps per Post	One
Size of Lamp	10,000—15,000 lumens
Spacing	90—115 ft.
Mounting Height . . .	15—18 ft.
Arrangement of Units	Opposite

FIGURE 5.9. Recommended lighting for major business streets in cities of populations over 20,000. From Ward Harrison, O. F. Haas, and Kirk M. Reid, *Street Lighting Practice* (New York: McGraw-Hill, 1930), 132–33. Reproduced with permission of The McGraw-Hill Companies.

the principal business streets and linked by underground wires—except downtown, where the business association had already subsidized "ornamental posts," each supporting a cluster of "white way" luminaires. On major residential thoroughfares, 100-candlepower lamps on posts alternated

C. Cities of 5,000—20,000 Population

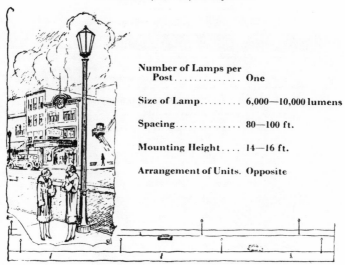

Number of Lamps per
Post.............. One

Size of Lamp......... 6,000—10,000 lumens

Spacing.............. 80—100 ft.

Mounting Height.... 14—16 ft.

Arrangement of Units. Opposite

D. Cities of Less Than 5,000 Population

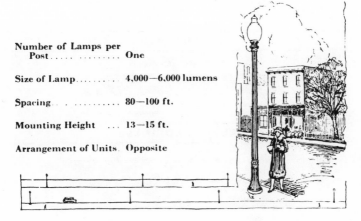

Number of Lamps per
Post.... .... One

Size of Lamp........ 4,000—6,000 lumens

Spacing.. . ... 80—100 ft.

Mounting Height ... 13—15 ft.

Arrangement of Units. Opposite

FIGURE 5.10. Recommended lighting for major business streets in cities of populations under 20,000. From Ward Harrison, O. F. Haas, and Kirk M. Reid, *Street Lighting Practice* (New York: McGraw-Hill, 1930), 132–33. Reproduced with permission of The McGraw-Hill Companies.

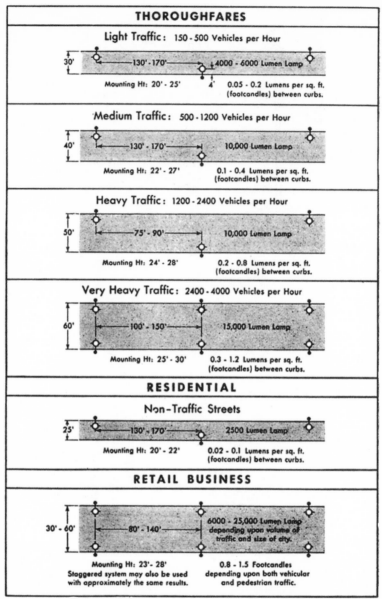

Recommended practice of street and highway lighting.

FIGURE 5.11. Westinghouse recommended lighting for urban streets as a function of traffic level. Residential and retail business streets were differentiated according to suggested mounting height and brightness of lamp. Thoroughfares with through traffic were similarly differentiated according to traffic volume. As such lighting standards were adopted nationwide, the look of urban streets at night came to display much uniformity city to city. From *Lighting Handbook* (Bloomfield, N.J.: Westinghouse, 1940); reprinted in "Chief Factors Involved in Street Lighting Plans," *American City* 59 (Jan. 1944): 87. Used with permission, Intertech Publishing Corp., © American City & County, Atlanta, Ga.

every 75 feet from one side of a street to the other, but on lesser streets the lamps, most suspended from power poles, were located only at intersections. Here, wires were strung from pole to pole.

In Indianapolis in 1926, five categories of street were defined for lighting purposes: primary business, secondary business, principal thoroughfare, secondary thoroughfare, and residential street. On primary business streets, lighting engineers used two 15,000-lumen Mazda C lamps per 20-foot post, with posts spaced on both sides of the street at 105-foot intervals. On secondary business streets, 15,000-lumen lamps were placed on 15-foot posts, similarly spaced. Principal thoroughfare streets had 6,000-lumen lamps placed on 15-foot posts, which were staggered every 110 to 125 feet and alternated from one side of a street to the other.[20] Globes of rippled glass with prismatic refractors varied in weight with each category of street.

After dark, distinctive light spaces emerged, encouraging and aiding automobile use in cities. Hierarchically arranged, these clear geographical variations could be noticed street to street in terms of light intensity, a circumstance we very much take for granted today. We expect main commercial streets to be the most brilliantly lit. Viewing a cityscape from above, we are not surprised to discern hierarchical relationships in light intensities across a darkened city's street grid. But in the 1920s and 1930s, when such a system was first introduced, its novelty offered further proof of lighting's essential modernity. A highly standardized visual order had been brought to the night.

HEIGHT, SPACING, AND POSITIONING

The widespread installation of arc lights in American cities brought with it the problem of glare. Early arc lamps, unshielded by the refracting glass of luminaires, were extremely harsh on the eyes. In replacing gas lamps on low posts, arc lamps proved unsatisfactory because of their blinding brilliance. To alleviate this harshness, arc lamps were raised high above city streets, either placed on very tall posts along street margins or suspended from cables well above street surfaces. The light tower enjoyed its brief popularity because it elevated the arc lamp even higher, well above the normal line of sight. Lamp height and the elimination of glare re-

FIGURE 5.12. Street lighting along Milwaukee Avenue, Chicago, 1998. Street lights blaze in the night, outlining a clear forward trajectory for auto movement. The brightness of the main street contrasts with darker nearby side streets, light defining the hierarchical relationship in the night.

mained an important preoccupation as lighting engineers sought to improve nighttime visibility, especially for motorists.

A motorist's ability to see objects silhouetted on a street or roadway was fundamental to lighting strategies. Glare substantially reduced such ability not only by directly obscuring objects ahead but also by reducing the contrast that made objects stand out against light reflected from the street surface. The discomfort glare caused was diminished by reducing luminance (or brightness), by increasing the reflective level of the road surface, and by increasing the mounting height of luminaires. Raising the height of luminaires from ten to thirty feet lowered glare by an order of thirteen, assuming candlepower, road surface, weather, and other conditions held constant.[21] With the introduction of luminaire reflectors and refractors, luminaire height came to influence, besides glare, the adequacy of light distribution on street and roadway surfaces.

Questions of recommended luminaire height, spacing, and positioning all revolved around the intensity of light produced by lamps, as mea-

FIGURE 5.13. Bird's-eye view west from Chicago's North Michigan Avenue, 1995. Streets coalesce in a grid of brightly lit patterns, outlining in the darkness Chicago's new River North entertainment zone.

sured in foot-candles or lumens. Uniformity of illumination, which was necessary for the continuous discernment of silhouetted objects while moving, was a function of light levels as measured midway between luminaires. In Milwaukee's early experiments with modern luminaires and posts, it was found that the optimum ratio of lamp spacing to mounting height was eight or less. "The spacing distance for any height of post," reported Milwaukee's lighting consultant, "can be readily obtained when the above ratios are decided upon, as 15-foot units spaced on a ratio of eight would be 120 feet apart, or 30-foot units spaced on a ratio of 8 would be 180 feet apart, etc."[22]

The positioning of luminaires had implications not just for uniform light distribution but for making movement safer. Where should the most intense light be located, and how should luminaires relate to one another in overall pattern? The most intense light, logically, needed to be located where automobiles turned corners or made other maneuvers, where traffic was heaviest, or where different modes of travel mixed. Street intersections were a logical choice. It was also sensible to arrange luminaires

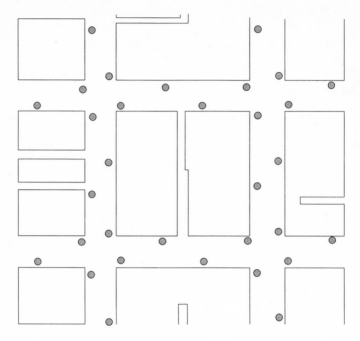

FIGURE 5.14. Street lighting for a downtown St. Louis block. Regular spacing of street-lights sets up a visual pattern, a frame of reference in the night for judging distance and anticipating movement ahead. From John E. Tiedman, "The New Down Town Street-Lighting," *Electrical Review and Western Electrician* 55 (Dec. 11, 1909): 1137.

in regular patterns against which one could judge distance and speed. A haphazard positioning of streetlights would be difficult to "read"—that is, it would not reinforce, especially at constant speed, a sense of steady progression. Conventions were developed for luminaire positioning, the first patterns more intuitive than based on objective measurement and analysis. In downtown St. Louis in 1909, four arc lamps were placed at each street intersection, with two at each alley entrance (fig. 5.14). Standards quickly produced a surprisingly limited array of configurations nationwide, even for various types of streets and highways: median, right-side, left-side, staggered, and opposite placements (fig. 5.15). Positioning of luminaires was determined by such factors as road design and direction and speed of traffic flow. In sorting out such factors, lighting engineers in-

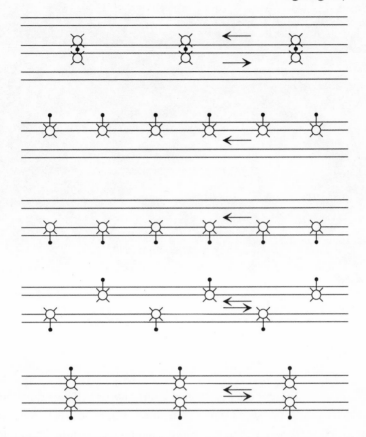

FIGURE 5.15. Luminaire mounting arrangements. The lighting engineer could choose from a variety of lamppost positioning schemes. Once chosen, the arrangement helped determine lamp and luminaire specifications. From *Roadway Lighting Handbook* (Washington, D.C.: U.S. Department of Transportation, 1978), 92.

creasingly turned to simple geometric configurations that were easily predictable in use.

AUTOMOBILE HEADLAMPS

Increasing automobile traffic on city streets and rural roads converted public ways into "machine spaces." By 1921, there were some 10.5 million motor vehicles registered in the United States, 8 million of them motor

FIGURE 5.16. A highway lit by headlamps. Illustrated is the intended effect of the then-innovative refracted lens. From an advertisement for the Warner-Lenz Company in *Cosmopolitan*, Sept. 1916, 167.

cars.[23] In 1924, there were 4.5 million more motor vehicles than there were electrically lit houses in the United States. Cars and trucks averaged six lamp replacements a year, including headlamps, taillights, and dashboard and other interior lights. Auto lamps of various sorts represented one-quarter of General Electric's total lamp sales, or some 36 million bulbs in 1924.[24] Headlamp technology evolved rapidly, driven, year to year, by the increasing speed of automobiles. First the kerosene lamp was adopted, fitted with a spherical reflector and glass cover. In 1904, acetylene lamps equipped with reflectors were introduced. In 1908, incandescent headlamps with tungsten filaments rated at thirteen candlepower appeared.[25] At night, automobile headlights and taillights in motion brought a sense of enhanced visual excitement to city streets.

Following development of the self-starter in 1912, which eliminated the need to hand-crank engines, generator-driven electrical systems were adopted, increasing voltage for headlamps. V-shaped coiled filaments were placed on the market in 1916, providing motorists with a more

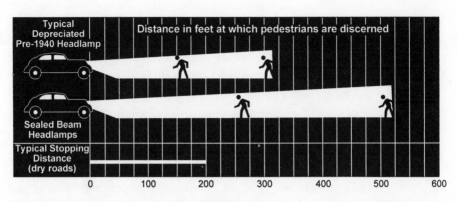

FIGURE 5.17. Visibility and stopping distances at forty miles per hour for automobiles with and without sealed-beam headlamps. From V. J. Roper, "The Amazing G-E Sealed Beam Headlamp," *Magazine of Light* 15 (Jan. 1946): 9.

concentrated light and increasing lumen output. At the same time, refracting lenses designed to diffuse light laterally appeared, which could better illuminate roadside margins and focus light downward, decreasing glare for oncoming motorists (fig. 5.16). The major visual challenges in night highway driving included seeing against the glare of oncoming automobiles, recovering from such glare once an oncoming car had passed, and adjusting to the lower levels of illumination that followed. Numerous devices were introduced to reduce glare by dimming, diffusing, or redirecting light, although most auto manufacturers merely tilted headlamps downward at an angle equal to half the spread of the light beam.[26] Headlamp glare was substantially reduced by General Electric's 1924 depressible-beam bulb, which carried two filament lamps of twenty-one candlepower each, one focused high and the other low, the precursor of today's "high-beam" headlights. A year later came the corrugated bulb lens, which broke up the shadows cast by the filaments.

Sealed-beam headlamps were made standard in the United States in 1940. They consisted of concentrated filaments at the focal point of a paraboloid of glass surfaced with a highly reflective metal and a glass lens cover with refracting prisms, both sealed together as a unit and filled with inert gas. Using only forty watts of electricity, the new lamps could illuminate road surfaces up to 1,000 feet ahead, a substantial improvement over earlier technologies (fig. 5.17).

STREET LIGHTING EFFECTIVENESS

The evaluation of street lighting systems was slow to develop. Just how much safer were well lit as opposed to poorly lit streets? Clearly, streets were more dangerous at night. One 1920 study showed that across the United States traffic accident rates were 38 percent higher at night, even under artificial illumination, than during daylight.[27] In 1935, more than 37,000 Americans were killed in automobile accidents, 22,000 of them at night. Between 1930 and 1935, daylight automobile-related deaths fell by 12 percent, but nighttime fatalities increased by 37 percent.[28] Nearly half of those killed at night were pedestrians hit by cars. Yet statistics collected seemed to suggest that, as street lighting improved, street movement became safer. After San Francisco increased lighting on Bay Shore Boulevard in 1933, the nighttime accident rate dropped by 40 percent in the first six months.[29]

More meaningful, perhaps, were data collected before and after decreases in street-lighting levels. Hit hard by the Depression, the city of Detroit dramatically reduced its street-lighting services in 1931. Lamps on alternating posts were darkened, taking over half of the incandescent filament lamps on main thoroughfares out of service. Nighttime traffic fatalities increased by 14 percent, while daytime fatalities decreased by 33 percent (the latter trend could be explained in part by the fact that there were fewer cars on Detroit's streets). The ratio of daytime to nighttime fatalities, 1 to 1 before the lighting cutbacks, increased to 1 to 2.4.[30] Beginning in 1933, Detroit street lighting was restored incrementally to its original levels, actually dropping nighttime auto-related fatalities below daytime levels.

How did improved lighting affect crime? Again, the statistics gathered were highly suggestive but inconclusive. Uneven sampling procedures from city to city made comparisons difficult. In Cleveland in 1930, 90 percent of the city's crime was committed after dark. Ninety-five percent of that crime was committed in the poorly lit areas outside the downtown "white way" district.[31] After Gary, Indiana, upgraded its entire street-lighting system between 1953 and 1955, the number of reported criminal assaults declined by more than 70 percent, and the number of robberies dropped by 60 percent.[32] Naturally, lamp manufacturers promoted light's crime-fighting implications. "A Policeman Every 50 Yards," head-lined one ad. "Efficient, modern street lights are like auxiliary policemen . . . Their revealing glow drives law-breakers away," it continued.[33] Per-haps, as the ad inadvertently admitted, lighting merely displaced crimi-nality, driving criminals to other areas that were still poorly lit.[34] It may also have displaced criminal activity from one kind of crime to another, reorienting law-breakers from crime on persons to crime on property, for example.

Conservation during the two World Wars offered opportunity to assess street lighting effectiveness. During World War I, street illumination in many cities was reduced as a coal-saving measure, so coal could be di-verted to uses more directly supportive of the war effort. Eastern seaboard cities, meanwhile, darkened their streets in order to thwart possible at-tacks from zeppelins.[35] During World War II, the threat from the air was very real for American cities on both coasts. By curtailing light, targets could be obscured and street-lighting patterns, useful to enemy pilots, could be diminished. In some cities Blackout-luminaires that concentrated light immediately below light poles were introduced. More widespread, however, were power "brownouts" and "dimdowns," in which lighting levels were substantially reduced, and "blackouts," in which lighting was completely screened. To aid motorists traveling in lowered light, adhe-sive reflective materials were used in the lettering of traffic signs, a prac-tice now commonplace. A special automobile headlamp was introduced that could not be seen from the air and yet emitted enough light to reveal large objects forty feet ahead.[36] Nevertheless, where such cutbacks were made, nighttime accident rates jumped.

As during World War I, accident rates soared during World War II when street lighting was curtailed. In New York City in 1941, the overall traffic death rate increased by 27 percent, but the nighttime rate jumped by 51 percent.[37] Nighttime accidents due to poor lighting caused an esti-mated loss of more than 200 million man-hours nationwide. Spurious sta-tistics were generated translating this loss into military equivalents. Ad-vocates claimed the lost manpower could have produced 2,000 heavy bombers or 6 battleships.[38] But by the end of 1943 it was evident that the country was secure from air attack, and the blackout drills and other pre-cautions were taken less seriously. A year later, most streets in American cities were again fully lit, although electric advertising signs would remain dark until the war's end.

TRAFFIC LIGHTS

Lights were used not only to illuminate traffic but also to regulate its flow. Traffic signals came to dominate major intersections in large cities, led by those on New York City's Fifth Avenue. In the center of intersections, at five block intervals, small houses were elevated on iron frameworks some eighteen feet above the pavement, each surmounted by sets of three lights—orange, red, and green. "When the orange light is burning, all traffic on the avenue moves and traffic on the cross-streets must stop," explained one reporter. "When the green signal is flashed, the reverse is true. Just before the signals are changed, the red lamp warns drivers . . . and gives them an opportunity to avoid . . . stopping too suddenly or being brought to a stop across the intersection of streets."[39] The signals were operated simultaneously from a master tower at Forty-second Street, and traffic policemen with hand-operated semaphores stationed at intervening intersections were cued by these lights. Detroit, Philadelphia, Baltimore, Milwaukee, San Francisco, and New Orleans all adopted such signal towers.[40]

The automated traffic signal followed as a means of reducing manpower and labor costs in traffic control. "Stoplights," as they came popularly to be called, began as small boxes containing three lamps—the now standard red, yellow, and green lights—visible through lenses on all four sides. Boxes with lights arrayed vertically were set on posts at the curb or hung on cables above street intersections. Popular also were two-sided boxes with a horizontal row of lights; these were fixed on brackets supported by posts (fig. 5.18). Along Broadway in New York City, from Rector Street to Eighty-sixth Street, horizontal-style traffic signals were used, one placed every four or five blocks downtown and every eight to ten blocks uptown.[41] Traffic moved for intervals of a minute and a half on Broadway and for forty-second intervals on the cross streets.

Traffic lights acting in concert created a novel visual effect, especially in combination with automobile headlights and taillights at night. A kind

FIGURE 5.18. Advertisement for the Crouse-Hinds Company. "Stop- and-go lights," railroad crossing flashers, and sundry other traffic lights signaled danger and cued movement. From *American City* 38 (1928), advertising section, courtesy of the Crouse-Hinds Co.

of ballet developed, as movement and cross movement was synchronized by green, yellow, and red. Viewed from above, the effect could be startling. But not everyone approved. The traffic lights irritated and depressed architect Charles Le Corbusier; visiting New York City in 1947, he wrote: "Terrible automatic red and green traffic lights, which bring ten thousand

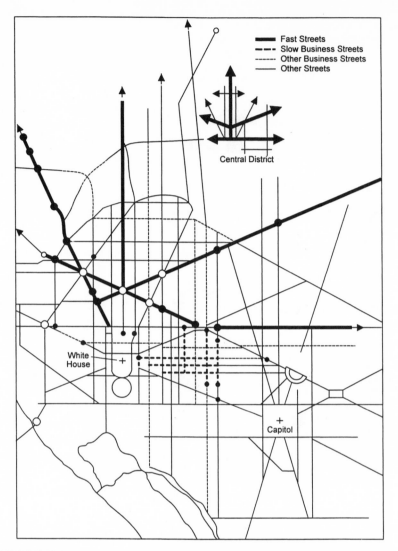

Fast Streets
Slow Business Streets
Other Business Streets
Other Streets

Central District

White House

Capitol

FIGURE 5.19. Plan for traffic signal coordination, Washington, D.C., 1930. Not only were major streets better lit, but coordinated traffic signals also kept traffic moving at steady speeds, another variable in the establishment of street hierarchies. From J. Rowland Bibbins, "High-Efficiency Signaling Raises Street Capacity," *American City* 42 (April 1930): 102. Used with permission, Intertech Publishing Corp., © American City & County, Atlanta, Ga.

yards of street traffic to a stop at one time and allow more than a hundred cross streets to function, extend their dictatorship over the whole metropolitan area and are an affliction to one's nerves."[42]

Traffic light technology improved steadily as more powerful incandescent filament lamps and more accurate reflectors and refracting lenses were introduced. Lights could be aimed, enabling them to better compete with electric advertising signs and other distractions night and day. Rules of thumb also developed: for example, on a two-way roadway where vehicle speeds exceeded fifty miles per hour, it was recommended that twelve-inch lights in a vertical plane be aimed two or three feet to the right of center line at a point some 1,200 to 1,500 feet in advance of the signal.[43] Turning arrows for turning traffic and "walk / don't walk" signs for pedestrians became commonplace after 1950. Unlike Europe, with its roundabouts and other like devices, Americans developed a substantial dependence on stoplights to regulate traffic.

The timing of traffic signals developed in the 1920s, expediting the through movement of traffic on major streets at set speeds. On Cleveland's Euclid Avenue, in 1924, traffic lights were installed at sixteen irregularly spaced intersections and timed such that motorists moving at a steady twenty-three miles per hour would not have to stop the entire length of the street.[44] Washington, D.C., installed a citywide system of coordinated traffic lights in the early 1930s (fig. 5.19). As they had done to determine street lighting, traffic engineers identified types of street by the character and volume of traffic—labeling business streets in Washington as express, fast, and slow, among other categories. The signals on each street were coordinated and "interlocked" with those of other streets. Light intensity was not the only thing that served to differentiate major thoroughfares from minor streets in a hierarchical geographical relationship, so too did the positioning and timing of traffic lights.

MOVING CARS AND TRUCKS very quickly came to dominate city streets in America. Street lighting ceased being about the creation of pedestrian places once mass production of motor vehicles began. Streets soon were for one purpose only—to move vehicular traffic as efficiently as possible. Street lighting was primarily intended to aid motorists. When pedestrians got in the way, as they so often did, especially at night, the call went out for brighter lights and more carefully engineered street lighting systems. It was the movement of machines that now dictated lighting levels in public places, thereby determining the general look of American cities at night. It is the look that we today readily accept.

The keys to brighter street illumination were the constant improvement of streetlight luminaires, the development of lighting systems designed to tie streets together hierarchically, the improvement of automobile headlights, and the coordination of improved traffic signals. A pervasive "traffic-orientation" was forcefully reinforced in night lighting. Light is, indeed, one of the great "form-givers" available to architects and city planners, but its application in the American city primarily came to serve the interests of traffic engineers. Most American city planning in the 1920s and 1930s focused almost solely on the accommodation of motor vehicles. "Getting people there" in rapidly moving automobiles, rather than "people being there" to enjoy diverse environments, became the important planning resolve.

Ironically, like all kinds of street improvement, more light in city streets often invited higher vehicular speeds, increased traffic flow, and inflated accident rates. Thus the promises made, indeed the basic assumptions brought to traffic engineering, locked the character of streets into an ever-tightening embrace of machine culture. Pedestrian control of public space diminished as light was intensified and spread more evenly across city space, favoring automobile travel. People moved rapidly through city streets while cocooned in metal and glass, turning less attention to public space and thereby inviting increased levels of criminality, even in well lit places. Street life, previously seen as a civic virtue, eroded in many urban locales.

CHAPTER SIX

Light as City Celebration

The lighting of public streets made them accessible for routinized night-time activity, but artificial illumination at night also encouraged special events that offered variation from the ordinary and the commonplace. From the early nineteenth century until well into the twentieth century—and even today—much of the excitement of nighttime rallies, parades, or outdoor theatricals derived from lighting itself. Special events provided opportunities to showcase light—to place light itself on display. Political rallies, festivals of various sorts, conventions, and the celebration of holidays, especially the Fourth of July and Christmas, contributed to these occasions. Celebratory light could give nighttime streets a carnival atmosphere, at least where street lighting did not overwhelm it.

Night was a time when the majority of Americans were not working, so artificial illumination allowed them to claim public spaces for recreation. For the more affluent, strolling or venturing out to restaurants, theaters, and pleasure gardens traditionally satisfied. For the less affluent, saloons and music halls provided venues. Meanwhile outdoor celebrations, which brought masses of people together on special occasions, mixed social strata. Indeed, such celebrations usually were staged for that purpose—to give the people of a city or town a sense of common purpose. Spectacular (or merely novel) illuminations at such events were an attraction to all. In the burgeoning industrial metropolises of the nineteenth century, light displays appealed to visual sensitivities across class lines. However, most light displays were carefully orchestrated to serve the goals of society's elites, who controlled lighting technology and who, of course, had the monetary wherewithal to subsidize its display.

Bright light signified sophistication and brought admiration for those who possessed it. A kind of social superiority was demonstrated, as the common person would be unable to duplicate the effects of spectacular lighting display. Those who raised the money, assembled the technical expertise, and administered successful light displays were admired and re-spected—in the public media, certainly—for their prowess at social or-ganization as well as their technical competency. A sense of accomplish-ment could not be denied. And when such accomplishment was turned to the celebration of a town or a city, or, indeed, the nation as a whole, on-lookers could be "electrified" by pride of place. In this way, artificial illu-mination was harnessed for social engineering.

NIGHTTIME POLITICS

Politicians in America were quick to embrace the outdoor rally. Rallies in streets and public squares proved an important mechanism for giving citizens a sense of participation in their new democracy. Artificial illumi-nation, especially from fireworks, was made central to Independence Day celebrations. Most of the nation had in fact ignored the Fourth of July until the mid-nineteenth century, when nighttime pyrotechnics achieved popularity. Political parties used brightly illuminated nighttime parades and other celebrations to garner votes in elections and to celebrate elec-tion victories. By the end of the nineteenth century, projected "lantern" light was used to report election returns in grand nighttime celebrations, the likes of which have not been equaled since.

Fireworks made from explosive nitrates were of Chinese invention; Eu-ropeans later used the nitrates as gunpowder for war. Lavish fireworks came into vogue in the Renaissance, and both Michelangelo and Leonardo da Vinci directed displays in sixteenth-century Italy. There was little color in fireworks until the nineteenth century, then Francois Marie Chartier, a retired artillery officer, published a pamphlet with formulae for various pyrotechnic colors in 1843.[1] Fireworks could be set off in containers on the ground, producing a shower of sparks resembling a fountain of water. They could be placed on scaffolding and connected by fuses to be fired in sequence. They could be shot high into the air either as canisters fired from mortars or as rockets. Aerial fireworks could be designed to "break" sequentially, releasing various combinations of color sent cascading to the ground.

Common names arose to identify various effects: Roman candles, geysers, peacock plumes, torrents, floating festoons. A stock repertory of displays were popularized in the United States, including the "Tree of Liberty," the "Goddess of Liberty," and shapes resembling the American eagle, the profile of George Washington, and the outline of the U.S. Capitol dome. Other less political pyrotechnic effects included the Egyptian pyramids, the aurora borealis, the casket of jewels, and Niagara Falls.[2] Fireworks programs usually climaxed with a large number of explosives fired at once, both on the ground and high in the air, offering a kind of visual pandemonium. Fireworks provided mass display. They could be witnessed by large numbers of people distributed over large areas.

Using engravings and detailed descriptions, *Harper's Weekly* and other illustrated magazines began to portray the Fourth of July celebrations held in New York City, popularizing the idea nationwide. In Manhattan, the Fourth of July fireworks display of 1859 reached its conclusion with the "Temple of Fame," a tall structure with three rounded arches. "Over the centre arch, resting on a . . . globe, was the national fowl of America, bearing an olive-branch in his talons, and contemplating with due gravity the word 'Union,' presented in blazing fires beneath him."[3] Below was the figure of George Washington in the act of delivering his "Farewell Address," while the other arches held the figures of "Liberty" and "Justice." A reporter confirmed: "Cheer upon cheer rent the air until the last spark had faded out." Flaming torches or flambeaux carried by the crowd helped light such celebrations, as did fixed gas lamps. Later, electric arc lamps contributed, but only with the coming of incandescent filament lamps did electricity play a truly decorative role. In 1914 for the Fourth of July, Manhattan was festooned with over 26,000 red, white, and blue bulbs strung on the trees in city squares and used to outline public buildings.[4]

Through the 1880s, until police began actively discouraging the practice, bonfires were lit in the streets of New York City on election nights— a practice some described as "the destructive instinct of Young Americans run wild."[5] Throughout history fires have been used for nighttime celebration, imbuing events with a degree of recklessness and lawlessness. Ash cans taken from private houses were filled with debris and lit, especially

FIGURE 6.1. The Wide-Awake Parade, New York City, Printing House Square, October 3, 1860. This parade, held to promote Abraham Lincoln's candidacy for the presidency, was one of New York City's first massive nighttime rallies, an event made possible by gas and other street lights, which enabled throngs of people to come and go throughout lower Manhattan after dark. From *Harper's Weekly*, Oct. 13, 1860, 648–49.

after saloons that had been closed for the polling reopened and drinking reduced crowd inhibition. Most political campaigns were organized around torchlight parades, which usually climaxed with outdoor assemblies for speech-making. In 1860, a "Wide Awake" parade was organized in support of Abraham Lincoln's bid for the presidency (fig. 6.1). One reporter described "thousands of torches flashing" in narrow streets "crowded with eager people," the windows of every house "swarming with human faces." The multitude of the street moved "as a vast river" with "the waving of banners and moving transparencies." Above it all fireworks exploded in every color. From Tenth Street to Union Square on Broadway, the "entire street sheeted with flickering light."[6]

The 1896 presidential election was memorable in New York City for the manner in which William McKinley's victory, and other election news, was reported. The city's newspapers competed by staging elaborate light shows. The *New York Journal* displayed election results in four locations. The side of the building near City Hall Park which housed the Journal's offices was covered with huge canvas sheets. A temporary four-story tower stood nearby, supporting a series of carbon arc projectors from which slide images were projected. Between news flashes, patriotic pictures were shown and a "projection microscope" was used to display the images of flies and other insects trapped between glass.[7] On the front facade of the building, a large "electric map" had been constructed, with flashing colored lights (green for Republican and red for Democratic) symbolizing results from state to state.

To the north, at Madison Square, the *Journal* erected two screens in addition to an electric sign of incandescent bulbs across which short, cryptic messages were electrically spelled out. There was a steam siren that gave out shrill screams whenever important returns were displayed. Balloons were tethered above the square, from which hung the portraits of important candidates, while searchlights on the ground illuminated them as relevant news was reported. Fireworks exploded overhead as winners were announced. The *Journal*'s other locations were the Brooklyn City Hall Plaza and the intersection of Madison Avenue and 125th Street. Some 25,000 paper megaphones, all inscribed with the newspaper's logo, were distributed at the various locations, and onlookers were encouraged to shout their approval or disapproval.[8]

The *New York World* used the side of its building for a canvas screen about 180 feet tall and sixty feet wide. Four projectors displayed returns. The *New York Times* similarly projected election results, both at its office building on Park Row and at Broadway and Twenty-third Street.[9] The *New York Herald* used the tower at Madison Square Garden, located just off Broadway, to announce results with color-coded searchlights. Along Broadway, theaters were brightly lit, many using their own slide projectors to share election news with theater-goers between acts. As one observer concluded: "Electric current was thought to be the proper thing for celebrating the election."[10]

FESTIVALS

Urban festivals were organized to dedicate new public facilities, to celebrate historical events significant to a city or to the nation, and to applaud military victories. With the rise of local political "machines" in American cities, society's elite tended to withdraw from politics, focusing instead on civic and cultural enterprises seen as uplifting. Parks, museums, libraries, hospitals, and universities attracted the philanthropy of the well-to-do, as well as their leadership. The organizing of festivals fit this trend. Celebrations in which entire city populations were invited to participate, if only as onlookers, held promise as community-strengthening, stabilizing events. One of the most extravagant celebrations in New York City's history marked the opening of the Brooklyn Bridge, in 1883, which tied the boroughs of two independent cities into a single giant metropolis. A massive fireworks display was presented on and from the bridge. Illuminated gas balloons rose above the bridge's span, and gas lamps lit its roadways.

Sometimes even new street lighting prompted public celebration. In San Francisco in 1916, an "Illumination Festival" was held to celebrate completion of the city's first ornamental street lighting system, installed from the Ferry Building west along a mile of Market Street, the city's principal thoroughfare. The luminaires and posts were taken from the site of the Panama-Pacific Exposition, held the previous year. The city's lavish new city hall, located in a new civic center, was also dedicated. A parade with electrically lit floats constructed on streetcar flatbeds told the story of artificial illumination. Thousands of costumed marchers, all carrying some form of light important in history, walked between the floats—the cave man with burning pine knot, the Assyrian with torch, the Greek with clay lamp, the Roman with candelabra.[11]

In October 1892, New York City celebrated Columbus Day with electrical "spectaculars," including a night parade with illuminated floats. An estimated $250,000 was spent on "signs, tableaux, and structural effects" across the metropolis. The Edison Building on Broad Street was outlined with greenish-blue electric arc lamps. The New York Life Tower was outlined by bands of red, white, and blue incandescent filament lamps. On that building's facade hung a mammoth portrait of Columbus framed by white and gold bulbs. One excited observer wrote: "Near the Metropolitan [hotel], on Niblo's Theater portico, a search light was perched which did remarkable execution as it flashed up and down the brilliantly decorated thoroughfare and revealed the great tide of humanity flowing between the solid banks of brick and stone mile after mile."[12]

In September and October of 1909, New York City sponsored the Hudson-Fulton Celebration. This gala was held in observation of the 300th anniversary of the discovery of the Hudson River by Henry Hudson and the centennial of Robert Fulton's steamboat. More than one million incandescent filament lamps and 10,000 arc lamps were installed, outlining streets and the bridges of the Hudson and East Rivers (fig. 6.2).[13] On Fifth Avenue between Fortieth and Forty-second Streets, a "Court of Honor" was formed by thirty-six large Corinthian columns, each topped by a golden globe. The columns were connected by a crisscrossing of light festoons making, according to one enthusiastic observer, "a grand electric canopy of such surpassing beauty that words may not describe it."[14] The six-mile parade route, from 110th Street and Eighth Avenue to Fourth Street and Fifth Avenue, was laced by double strings of incandescent lamps.

At Riverside Drive and 155th Street crowds could view an electric "scintillator," developed by W. D'Arcy Ryan, one of the nation's leading designers of light spectaculars. "Forty huge searchlights of varying color shot enormous beams high in the air, now radiating in fan-like effect and changing from intensest white to the softer greens and yellows; now again shifting bodily from east to west and back again with frightful speed."[15] Clouds of steam were vented up to enhance the atmospheric effect of the light (fig. 6.3). Colors were changed with filters.

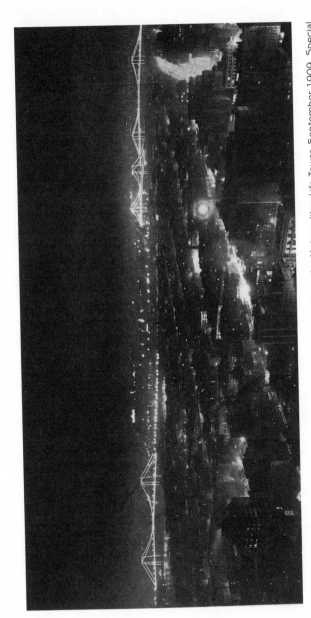

FIGURE 6.2. Lights of the Hudson-Fulton Celebration as viewed from the Metropolitan Life Tower, September 1909. Special festoons of electric incandescent filament lamps outlined bridges and landmark buildings and lined major thoroughfares. The city, when viewed from on high, seemed painted in white light. From *Thirty Years of New York, 1882–1912* (New York: New York Edison Co., 1913), 135.

FIGURE 6.3. The "Scintillator" of the Hudson-Fulton Celebration, September 1909. Light from giant searchlights was refracted through steam to produce a visual spectacle. From *Thirty Years of New York, 1882–1912* (New York: New York Edison Co., 1913), 259.

At the end of the Spanish-American War, Chicago held a "Peace Jubilee," featuring "myriads of electric lamps . . . stretched canopy-like across the streets."[16] At the end of World War I, the city created a "Victory Way" in Grant Park on its lakefront, anchored by an "Altar of Victory." Between two ninety-foot columns resembling candelabra was suspended a "Curtain of Jewels," made up of tens of thousands of small glass prisms. Hung on the curtain was the outline of a large American eagle bearing the flags of the allied nations. The whole was floodlit. Red and orange lights were concealed in the tops of the columns, giving each the effect of a lit torch, an impression enhanced by the release of steam into the air.

CONVENTIONS

Throughout the nineteenth and twentieth centuries, fraternal organizations, trade and professional associations, and veteran and church groups with national and even international memberships have held large annual conventions for which cities vigorously competed. It was the mature railroad network of the country that enabled delegates from around the nation to gather in large assemblies numbering in the thousands. Small town Americans especially were great supporters of the Masons, the Elks, the Rotary, and other organizations in which an element of gregariousness was played out in provincial rituals and in which members could also claim connection to cosmopolitan big city life. Annual convention trips to big cities loomed large in members' imaginations as chances to broaden life's experiences.

Plans for spectacular illuminations were important in attracting conventions to a certain city. The bright lights of the metropolis beckoned deliberately. Much of the visiting member's sense of fulfillment came from the special attention bestowed on his or her organization through spectacular illumination. Such lights validated visits, and visitors, as special. Not only being in a city but being there at a special time—the message that spectacular lighting sent—added substantially to conventioneer satisfaction. Remembering the place and sharing those memories with others was made all the more vivid by the festive lights. Thus lighting technology was harnessed to enhance visitors' sense of time and place, adding to their self-fulfillment.

In 1891, the Grand Army of the Republic met in Detroit. At the river-front, welcoming delegates arriving both by lake steamer and by train, was a forty-eight-foot electric sign depicting an American flag, an eagle, a cannon, a horse's head, and an anchor. "HAIL VICTORIOUS ARMY" and "G.A.R." were written in tall letters. More than 2,000 incandescent bulbs tinted red, white, blue, and yellow were used; their glow, it was claimed, was visible for ten miles.[17] In 1892, the G.A.R. met at Washington, D.C., where red, white, and blue lights decorated Pennsylvania Avenue. At both ends of the avenue eighteen-foot signs each carried an eagle, a cannon, and the stars and stripes lit in appropriate colors. Government buildings were specially floodlit, and on the White House lawn, wrote one observer, "every bush, tree, fern, and flower bed glistened with the light of innumerable miniature lamps."[18] On the roof of the Washington Loan and Trust Company, swiveling search lights created moving beams of light. At one viewing stand the name "Lincoln" was spelled out in white incandescent bulbs.

Perhaps the various Masonic orders under the Knights Templars, which met every third year, generated the most lavish light displays. In Pittsburgh in 1898, the city set out twenty-nine electric arches, which, by one account, "threw a radiance along the streets that gave one the impression that the atmosphere had been transformed into a sea of incandescence."[19] On the tower of the Allegheny Courthouse, a 100-foot cross was outlined in red and white lights. Similar displays were created on the facades of other tall buildings, employing an estimated 50,000 bulbs across the city's downtown areas. In San Francisco in 1904, the outline of a huge bell was suspended one hundred feet above the intersection of Market and Third Streets between three newspaper buildings—those of the *Call*, the *Chronicle*, and the *Examiner*.[20] On the tower of the Ferry Building, at the foot of Market Street, a giant Maltese cross was outlined in red lights. Just outside the railroad station a large welcome arch was erected, covered with incandescent bulbs. In Denver in 1913, the welcome arch was in the shape of a giant knight mounted on horseback and lit by search lights placed on the Foster Building some two blocks away (fig. 6.4). Such lighting effects were celebrated in postcard art specially prepared for conventioneers.

FIGURE 6.4. Postcard View of the "Mounted Knight Crusader." Thirty-Second Triennnial Conclave, Knights Templar, Denver, 1913. The card was marketed for sale to conventioneers, a souvenir of a special time and place.

ARCHES AND LIGHT FESTOONS

If lamps on posts marginal to streets created exciting, rhythmic lighting effects, then the use of arches and other light festoons amplified that excitement severalfold. Light festoons were possible following the introduction of electric incandescent filament lamps, which could be strung in linear series on cables from light poles (fig. 6.5). They could also be arranged on sturdier arches, metal frameworks specially erected as light supports (fig. 6.6). Such installations were made initially in cities like Columbus, Norfolk, Portland in Oregon, and Denver, as festival lighting, not intended to be permanent. The visual effect proved so striking and attracted so much pedestrian traffic downtown, however, that cities, encouraged by merchant associations, began to maintain arches year round. Denver, prompted by its Chamber of Commerce and the Denver Gas and Electric Company, determined to make itself the nation's "City of Lights." Not only were its arches retained, but a giant "Welcome Arch" was also constructed just outside the Union Depot. The first thing railroad travel-

FIGURE 6.5. View of electric incandescent filament lamps festooned along Pennsylvania Avenue, Washington, D.C., circa 1920. Light displays helped conventioneers establish their time in a place as special. Municipalities created elaborate lighting proposals in bidding for convention business.

ers saw on entering the city's downtown district at night was a lavish light display.[21]

Strollers and motorists on streets outlined with arches of incandescent filament lamps enjoyed a carnival-like glow. Especially at night, this sense of perspective was accentuated as people looked up and down the festooned streets. Critics pointed out that during the day the arches represented a kind of clutter, somewhat like overhead wires. Indeed, in Norfolk the arches added substantially to the confusion of power poles and wires overhead. When several arches appeared on a block, they created a "shed-like" effect "contrary to the feeling of openness and freedom which [a] street should possess," wrote one commentator.[22] As another critic ridiculed: "The so-called 'ornamental' lighting consisting of luminous arches and festoons, draped over business thoroughfares, scarcely deserves a word in passing. Its suggestion savors of the country fair, the temporary fete, and the lighting of the boardwalk or promenade at the seaside resort."[23] The skeleton frameworks, he asserted, gave business districts "the air of a dismantled factory." Perhaps raising even more criticism was electric arch interference with display window lighting and with electric signs

FIGURE 6.6. "Electric Arches" on High Street, Columbus, Ohio, circa 1910. So popular were these radiant light festoons that many cities moved to incorporate them into permanent downtown street-lighting schemes.

on stores. In addition, arch lighting was expensive to maintain, individual bulbs continually requiring replacement.

Light festoons were also used to outline buildings, although few businesses could afford the extravagance, except for the utility companies that generated the power. One of the most elaborately lit was the Gas and Electric Building in Denver (fig. 6.7). The structure was promoted, with typically American exuberance, as "the Best Lighted Building in the World." Its tripartite design was emphasized—the building's base, shaft, and capital each outlined in light differently. This early skyscraper stood in the night as a brilliantly lit advertising sign for electricity and its lighting capabilities.

SEASONAL HOLIDAYS

Over time, seasonal festivals have been promoted in most American cities, at least intermittently. None, perhaps, has proven so enduring as New Orleans' Mardi Gras, in which nighttime parades and specially lit

DENVER, COLORADO

17170

FIGURE 6.7. The Gas and Electric Building, Denver, circa 1910. Encouraged by its largest public utilities company, Denver sought to establish itself as the nation's "Capital of Night Lighting." Light festivals complete with light parades were promoted in many cities.

parade routes figure prominently. Portland, Oregon, still puts on its summer Rose Festival today, but Denver's fall "Festival of Mountain and Plain" proved short-lived. Both used extravagant lighting to attract out-of-town visitors, and in both cities, "electric clubs" helped organize

FIGURE 6.8. View off Soldiers' and Sailors' Monument in downtown Indianapolis, 1992. Festooned lights make the monument appear to be a giant Christmas tree. Perhaps no other holiday light display sponsored by a city government was ever quite so spectacular.

events.[24] But no season has proven more conducive to the use of special lighting effects in the United States than Christmas and Hanukkah. In downtown areas and main streets nearly everywhere, December has become a month of special lighting.

Through the early nineteenth century, relatively little celebration was made of Christmas beyond the family and the church. With the rise of the department store after the Civil War, itself a symbol of growing American affluence, the giving of gifts and the staging of elaborate dinners in wealthier households prompted a short shopping season immediately before Christmas Day. On Christmas Eve 1882 in New York City, Broadway, Fourth Avenue, and Fifth Avenue were set ablaze with electric arc lights. The stores were all open and thronged with buyers. The Washington Market seemed to James McCabe "the very incarnation of Christmas."[25] "The scene especially at night, almost baffles description," he wrote. "Long rows of turkeys hang from the hooks of the stalls which usually groan beneath the weight of butcher's meats and sugar-coated hams. Wreaths and festoons of evergreens, mingled with holly-berries, decorate every stall, and the great sheds are aglow with hundreds of lamps of every description."[26]

In 1912, New York City erected in Madison Square a "Tree of Light" decorated with colored incandescent bulbs. News stories sent out over various telegraph news services elicited inquires from hundreds of interested towns and cities. Marking the Christmas tree's seasonal location in New York City, a permanent "Star of Hope" was placed "to lighten somewhat the increased gloom that seems to settle down over the city when the Tree of Light is gone."[27] In the 1930s, New York City's symbolic Christmas tree was moved to Rockefeller Center. Safe, inexpensive strings of incandescent filament tree lights flooded the market for domestic holiday use after World War I, exciting even more public interest. City after city began to decorate a ceremonial tree and festoon incandescent colored light (usually subsidized by merchants) along commercial streets. In 1933 4 million sets of Christmas lights (eight bulbs were standard to a set) were sold in the United States, roughly 10 percent of them for outdoor display.[28] In total, 90 million Christmas bulbs were believed to be in use. "Apartment houses, private dwellings, stores, garages, even filling stations express the holiday spirit with every form of outlining, window-blazing, wreathing and festooning."[29]

Use of decorative Christmas lighting surged in the 1930s, with many localities sponsoring lighting contests. In Des Moines, the intent was "to create a good old fashioned Christmas spirit by getting people to decorate their homes and lawns."[30] In California, the cities of the San Joaquin Valley vied with one another in a highly publicized Christmas lighting competition. Lighting enthusiast Clarence Pratt of San Francisco organized the Outdoor Christmas Tree Association to promote such outdoor display. Movie star Mary Pickford associated herself with the group, promoting outdoor Christmas decorating in Southern California.[31] Other cities were soon creating light festivals at Christmas. At Bethlehem, Pennsylvania, in 1948, Christmas decorating focused on the "Hill to Hill" bridge, connecting the two principal parts of the city over the Lehigh River. "Streamers" of red and green bulbs outlined the bridge, and a giant Christmas tree was erected at the bridge's downtown approach. On the city's main commercial street, clear glass luminary globes were replaced by globes of green glass, while red and green light festoons formed a canopy over the street. Above the city on South Mountain a giant electric "Star of Bethlehem" was lit.[32]

Lamp technology favoring larger and more durable Christmas lights continued to evolve. But in 1960, strings of minute white lights, called Silvestri bulbs, after their inventor George Silvestri of Milan, Italy, appeared. Chicago's North Michigan Avenue—the city's "Magnificent Mile"—was the first American street lit with these new lamps.[33] The Greater North Michigan Avenue Association subsidized their installation both on trees lining the boulevard and on the Water Tower, the castellated Gothic landmark that had survived the disastrous 1871 fire and from which electric arc lamps had first been demonstrated in the city.

New Year's Day falls shortly after Christmas and, indeed, marks the end of the traditional "holiday season," as Thanksgiving marks its beginning. Many cities came to sponsor New Year's Eve celebrations, although interest dwindled rapidly after World War I. In San Francisco, both the Ferry Terminal, with its clock tower, and Union Square were the centers of celebration. In 1909, the streetlights of downtown were slowly dimmed and then darkened completely, to be brightly relit at the stroke of midnight. "Suddenly the air was filled with light," wrote a spectator. "Market Street, Union Square, the bold facade of the St. Francis hotel and the decorated skyscrapers along the great thoroughfare of the city, gave forth millions of rays of incandescence." He concluded: "The carnival spirit was rife and under the glittering sheets of light, between the rows of gaily bedecked skyscrapers, the throng danced and shouted."[34] From 1920 onward, New York City's Times Square would claim the greatest attention on New Year's Eve. There the electric "time ball" atop the New York Times Building lowered to mark the start of each new year, a scene communicated nationwide in movie newsreels and later on live television.

THE USE OF LIGHT AS AN ELEMENT of celebration spread rapidly in the United States as new lighting technologies emerged. In many celebrations, especially in the nation's largest cities, light and lighting often became the principal focus of such festivities. Although it was also staged in other ways, New York City's Hudson-Fulton Celebration of 1909 was known primarily for its lighting spectaculars (and today is remembered for little else). Giant electric signs, light-festooned streets, and brightly illuminated buildings at festival time became important urban icons, coloring how people thought of cities. Much visual imagery of urban America formed around the exaggerated impressions of festival lighting. City streets were made rather like theater stages, on which the drama of life was made to play out as spectacle.

Prime among those theatricals were the celebrations of American politics. Large crowds could be assembled only after daytime work hours. The night atmosphere lit in a dramatic way could serve to amplify the importance of a political message while offering entertainment, demonstrating that a sense of circuslike diversion was never distant in American politics. Electric lighting allowed political rituals to be acted out in large public places, engaging a wide range of people and contributing to a sense of widespread public participation in American democracy. With electric light, civic ceremonial spaces could be used to fuller advantage, light used to direct and focus attention and even choreograph parades and other rituals. Today, however, the political rally has become a media event staged largely for a television audience. It can be held at any time and recorded, edited, and played back at will; outdoor lighting has been made largely irrelevant in the modern political rally, save its use in preparing a broadcast.

Light celebrations also came to stand as measures of seasonality. Today, the most widespread use of festive lighting celebrates an end-of-year holiday season built largely around Christmas. With their own Christmas lights, both indoors and out, ordinary Americans play roles previously reserved for the lighting engineers. They manipulate light for highly personal reasons, albeit in highly standardized ways. Festive lighting, which started in public outdoor spectaculars to define community, has now become more a means of private business and household expression. Christmas lighting that serves commercial purposes is fully intertwined with electric advertising art. Public festive lighting came to seem overly commonplace, and use of light in public celebration lost impact because city lights were overwhelmed by the brilliance of streetlights. Even before World War I, streetlights were necessarily dimmed or extinguished temporarily in order to amplify the visual power of festive light. After World War II, traffic needs totally precluded such romantic pretense.

CHAPTER SEVEN

Lighting the World's Fairs

While outdoor celebrations were used to draw people together as urban communities, the world's fairs did that on a much grander scale. Beginning in the 1850s, lavish expositions were organized to celebrate national pride, and various cities competed to out-stage one another in demonstrating America's technological and artistic achievements. The intention was to give all Americans a feeling of participating in a national experience superior to all others, the fairs serving to establish America and Americans as special. Like other forms of public celebration, the fairs were designed to set social agendas as well. At a time of increasing social diversity and class stratification, the fairs advanced gentry, or elitist, cultural values, promoting, in other words, the social status quo. However, the fairs also showcased some more vulgar popular entertainments that went beyond elitist notions of what was tasteful and uplifting—pointing, in the process, to a future of change.

The world's fairs were instrumental in introducing Americans to nighttime lighting. New lamp technologies and new lamp applications were often first seen at these large expositions, either in promotional displays or actually used in the lighting of buildings and grounds. Applied at the fairs, however, night lighting was substantially divorced from the functional needs of ordinary city streets. There, specifically, the automobile did not have to be accommodated. Consequently, the world's fairs misled Americans, giving them false notions of how their future cities would look at night. The preconception was encouraged that celebratory and informative uses of night light would enjoy fuller application.

Lighting technology would prove the major attraction at many fairs, and not always by deliberate forethought of design. At night, lighting could amplify the grand and the monumental. "Courts of honor" featuring colonnaded exhibition halls, terraced gardens, and lagoons and fountains, were lit at night—striking a balance, in the parlance of the time, between the sedate and the sublime. Light also could be used for garish display along the "midways," highlighting their freak shows, music halls, carnival rides, and other exotica. Ultimately, the sedate and the vulgar were merged in world's fairs, which did not celebrate past achievement using historical architectural referents so much as they imagined future possibility using tomorrow's technology. Promoters sought to use night light innovatively, either by adopting a new lighting technology to demonstrate it for the first time or by applying old technologies in truly novel ways. The story of lighting at world's fairs, therefore, necessarily proceeds chronologically, with a focus on the important firsts touted by promoters who sought to outdo one another in the "packaging" of nighttime spectacle.

The word *fair* derives from the Old English *feire,* meaning *holiday.*[1] It was often used to describe feast days sanctioned by the church, times of financial security that attracted merchants. Commercial bazaars typically developed around churches on feast days. In the nineteenth century, the words *exhibition* and *exposition* were used to describe the formal showcasing of industrial technology in special pavilions constructed for national and even international audiences. Lesser expositions, usually dubbed "fairs," were more local or regional in importance and were often conducted on an annual cycle—county and state fairs, for example. When expositions grew into mammoth undertakings spread across massive complexes of buildings located in extensive parklike settings, the "world's fair" was born. Between 1853 and 1968 there were in the United States at least nineteen major and twenty lesser expositions (see table 7.1).

THE EARLY FAIRS

The first major American fair was New York City's 1853 Crystal Palace Exhibition, patterned after London's successful Crystal Palace, which had opened two years earlier. Occupying a single exhibition hall constructed of iron and glass and located at Broadway and Forty-second Street, the

TABLE 7.1. Major Expositions in the United States, 1855 to 1968

City	Date	Name	Attendance[a]
New York City	1853	Crystal Palace Exhibition	1,250,000
Philadelphia	1876	Centennial Exhibition	8,004,274
Louisville	1883	Southern Exposition	
New Orleans	1884	World's Industrial and Cotton Exposition	
Chicago	1893	World's Columbian Exposition	21,477,212
Omaha	1898	Trans-Mississippi and International Exposition[b]	
Buffalo	1901	Pan-American Exposition	8,304,073
St. Louis	1904	Louisiana Purchase Exposition	12,804,000
Portland	1905	Lewis and Clark Centennial and American Pacific Exposition and Oriental Fair	
Seattle	1909	Alaska-Yukon-Pacific Exposition	
San Francisco	1915	Panama-Pacific International Exposition	13,127,103
San Diego	1915	Panama-California Exposition	
Philadelphia	1926	Sesquicentennial Exposition	5,852,783
Chicago	1933	Century of Progress Exposition	39,052,236
San Francisco	1939	Golden Gate International Exposition	
New York City	1939	World of Tomorrow Exposition	44,932,978
Seattle	1962	Century 21	9,639,967
New York City	1964	New York World's Fair	51,607,548
San Antonio	1968	Hemisfair	6,400,000

[a] Reid Badger, *The Great American Fair: The World's Columbian Exposition and American Culture* (Chicago: Nelson Hall, 1979), 131.
[b] Continued in 1899 as the Greater American Exposition.

fair was managed, to the delight of over a million visitors, by none other than Phineas T. Barnum. Although the show's intent was serious, lacking the vulgarity and hokum of Barnum's American Museum, there was one important crowd-pleasing entertainment. A steam elevator lifted visitors to an observation platform 300 feet above the exhibition floor. At night the scene was especially sensational, there being more gas lights in the exhibition building than on all the streets of the city.[2]

The floor of New York City's Crystal Palace was organized rather like a bazaar, with machines and other objects placed on display in small booths and Victorian bric-a-brac filling showcases. The intent was to illustrate human progress in an opulent display of nineteenth-century materialism. The building itself was little more than "an exaggerated greenhouse," although its sheer size made it impressive.[3] Its construction was based on previous advances made in railroad bridge building, an engineering art at which Americans excelled. Although intended as a permanent exhibition hall, the short-lived structure was consumed by a spectacular nighttime fire in 1858.

New York City's Crystal Palace was followed by a series of fairs held in European cities: Paris in 1855 (and again in 1867), London in 1862, and Vienna in 1873. At those fairs, several exhibition halls were filled and displays were built showcasing not only technology and commercial products but also "cultural" artifacts, especially art symbolic of European superiority in taste and fashion. Soon the scene was set for another American exhibition, a celebration of the nation's Centennial at Philadelphia.

Like its European counterparts, the 1876 Centennial Exhibition was laid out in an extensive pleasure garden. Formal vistas, grand avenues, giant fountains, and extensive terraces were created. Exhibition halls were grouped as an ensemble, the whole made to seem greater than the sum of the parts. It was intended to appear as an ordered world, similar to European palace complexes set in landscaped parks. The exhibition was suggestive of an urbanity associated with great European cities—it possessed an axial formality, a clear sense of center, and classical symmetry.[4] It was an authoritarian plan that was more a tribute, perhaps, to aristocratic power and social hierarchy than to democratic and republican ideals. A giant observation tower overlooked the grounds, served by two steam elevators. Philadelphia's was the last great fair to use the steam engine as symbol of progress. Henceforth, the electric dynamo and electricity would take center stage.

Elitist if not aristocratic tastes dominated inside the Philadelphia fairgrounds. Beyond the main gate, however, there spread an extensive entertainment zone, largely unplanned and uncontrolled by the exhibition managers. Popularly called "Dinkeyville" or "Shantytown," it comprised small hotels and boarding houses, restaurants, beer gardens and saloons, and a wide array of carnival entertainments very definitely of the ilk P. T. Barnum represented. Whereas visitors were expected to display a serious, informed, and highly disciplined decorum inside the grounds, outside the

TABLE 7.2. Names of Entertainment Zones at Selected World's Fairs

City	Date	Name
Philadelphia	1876	Dinkeyville (also Shantytown)
Chicago	1893	The Midway Plaissance
Atlanta	1895	Midway Heights
Nashville	1897	Vanity Fair
Omaha	1898	The Midway
Buffalo	1901	The Pan
St. Louis	1904	The Pike
Portland	1905	The Trail
Seattle	1905	The Pay Streak
Jamestown	1907	The Warpath
San Francisco	1915	The Joy Zone
San Diego	1915	The Isthmus
Chicago	1933	The Midway
San Francisco	1939	The Gay Way
New York City	1939	The Great White Way
Seattle	1962	The Gay Way
New York City	1964	Amusement Area

gate people were allowed to express various degrees of vulgarity in undisciplined irreverence. Rather than be "improved" there, visitors could merely have fun. And much of that fun was had at night in the glow of gas light.

The word *midway* became the generic descriptor for these entertainment zones at the large expositions, although other terms were actually used in most instances—*Pan, Pike, Trail, Pay Streak, War Path, Joy Zone, Isthmus, Gay Way* (see table 7.2). After Philadelphia's fair, the midway entertainments were planned as adjuncts to the main exhibits, but they were still held at some distance from them. With time, midways were given more central locations as differences between "high brow" and "low brow" tastes broke down. Ultimately, the courts of honor and the midways would be substantially blended in design and function.

Louisville's Southern Exposition in 1883 was the first to embrace electric lighting for illumination. It was also the last to use gas lamps. An Edison system with 4,000 sixteen-candlepower incandescent filament lamps was installed, with light festoons outlining the exhibition halls and electric arc lamps on ornamental posts lighting the grounds. Companies competed aggressively for lighting contracts at the fairs, and this one was the largest electrical lighting contract negotiated to that time. General Electric saw the Louisville fair as an extraordinary opportunity to demonstrate its wares (companies also competed aggressively through lavish exhibits). The General Electric display at Louisville included both a "tower of light," sheathed in incandescent filament bulbs and prism glass, and a large electrically illuminated fountain.

Because of the exposition, the City of Louisville claimed title to "the brightest spot on earth."[5] In historian David Nye's words, for the promoters of such fairs, spectacular lighting proved to be "dramatic, non-utilitarian, abstract, and universalizing." It provided at night "a brilliant canopy, connecting the many exhibits, statues, fountains, and pools in one design that was at once refined, ethereal, and stunning."[6] Technically sophisticated electrical lighting displays brought prestige to expositions. And so, at the "World's Industrial and Cotton Centennial" at New Orleans in 1884, electricity was necessarily employed, this time with a tower system for outdoor arc lights in which five towers each held fifty 2,000-candlepower lamps.[7]

THE WORLD'S COLUMBIAN EXPOSITION

Honoring the 400th anniversary of Christopher Columbus' first voyage of discovery across the Atlantic, the 1893 World's Columbian Exposition was also an opportunity for the City of Chicago to celebrate its rebirth following the disastrous 1871 fire. As chief of construction for the fair, architect Daniel Burnham coordinated the work of many of America's leading artists and designers, such as landscape architect Frederick Law Olmsted, sculptors Daniel Chester French and Augustus Saint-Gaudens, muralists John LaFarge and Elihu Vedder, and the principals of such architectural firms as McKim, Mead, and White and Adler and Sullivan. Consensus on a unified plan of neoclassical ornamentation was achieved easily, as most of the design participants had received their formal training at the Ecoles des Beaux-Arts in Paris, an advocate of "respect for tradition, proportion, symmetry, monumentality, grandeur, and vista."[8] The result was a monumental "White City" in the image of Rome or, given the extensive use of water in reflecting ponds, Venice.

The fairgrounds were intended to demonstrate what cities could be and, equally important, that Americans were capable of urban vision despite the actual commercial vulgarity and utilitarian sameness of cities like host Chicago. At issue was the hegemony of elite values in the creation of a future urban world, values seemingly threatened in the increasing economic dislocation and social and political strife of the new industrial era. Ideas of cultural unity, public order, and civic virtue were central in the fair planners' message. With that in mind, they created "a magnificent stage prop set in a landscape of fantasy."[9] Critics, however, saw the fairground as an insult to American design genius. They saw ingenuity more clearly in Chicago's many new office towers. Nonetheless, fairs were not reflections of real cities—they were intended be purveyors of fantasy. They were displays intended to take people out of their ordinary routines, removing them, if only temporarily, from their usual positions in society's stratification.[10] Thus removed, people were susceptible to messages both overt and covert. Value preachment seemed to better fit the "timelessness" of classical grandeur rather than the utility of nascent modern architecture.

A stunning ensemble of buildings was organized around a formal basin nearly a mile long, the surrounding exhibition halls all connected by colonnades (fig. 7.1). At the western end loomed Richard Morris Hunt's gold-domed Administration Building, recalling the Duomo in Florence. At the eastern end stood Daniel Chester French's giant statue, "The Republic," a female form dressed in a Grecian toga and holding aloft an eagle perched on a globe. Farther east, separating the basin from Lake Michigan, was the "Peristyle," a series of forty-eight Corinthian columns—one for every American state and territory—with a triumphal arch at center topped by a statue of Columbus standing in a horse-drawn chariot. To the north of the Court of Honor was a series of informal basins, the "Wooded Island" with its curved walks contained in the largest. Here picturesque romanticism offered a sharp contrast to the classical monumentality nearby. Due west of the Womens' Building was the "Midway Plaissance," the entertainment zone (the French word for *pleasure* tacked on for a pretentious effect).

FIGURE 7.1. The Court of Honor at night, at the World's Columbian Exposition, Chicago, 1893. Many visitors by day found the "White City" overwhelming given its vast size. But at night the dazzling lights edited the visual stimulae, reducing the scene to a more comprehensible scale. From J. P. Barrett, *Electricity at the Columbian Exposition* (Chicago: R. R. Donnelley and Sons, 1894), n.p.

FIGURE 7.2. An electric arc lamp silhouetted against an illuminated fountain at the World's Columbian Exposition, Chicago, 1893. From J. P. Barrett, *Electricity at the Columbian Exposition* (Chicago: R. R. Donnelley and Sons, 1894), n.p.

At night the fair was lit by electric arc lamps on ornamental posts, arranged to reinforce either the geometry of the formal plazas or the informality of the curvilinear walks (fig. 7.2). Arc lamps were also used in the "electric fountains" and in the searchlights that both floodlit statuary and sent light beams up into the night sky, the work of lighting de-

FIGURE 7.3. Searchlight at the World's Columbian Exposition, Chicago, 1893. Flood-lighting, introduced at the world's fairs, was widely applied beyond to the lighting of city landmark structures, especially tall office buildings. From J. P. Barrett, *Electricity at the Columbian Exposition* (Chicago: R. R. Donnelley and Sons, 1894), n.p.

signer Luther Stieringer (fig. 7.3). Clusters of electric arc lamps, suspended from above, lit the floors of the exhibition halls. Incandescent lamps supplemented the arc lamps in the Electricity Building, making it the only truly well-lit exhibition hall of the fair. In total, an estimated 92,600 electric lamps of various descriptions were used to light the buildings and the grounds.[11] The fair's searchlights alone used three times the amount of electricity then employed to light Chicago streets.[12] At night, crowds were drawn to see General Electric's new "tower of light" (enlarged and improved over the one in Louisville in 1883) in the Electricity Building, a column over 80 feet tall and studded with 10,000 incandescent filament lamps reflected and refracted with mirrors and glass prisms. Yet it was a Westinghouse system that actually lit the fair, the Westinghouse Company having seized the opportunity to demonstrate its alternating current technology.

Electric light made the White City gleam in the night. Popular consensus held that the fair was best experienced after dark, so dazzling was the visual display. "He who has not seen the World's Fair at night has not seen it at all," claimed a reporter for the *New York Times*. He continued:

> As the darkness deepen[s], the blue of the sky behind the colonnades of the peristyle [becomes] a dusky gray. Suddenly a serried row of lights appears to mark the cornice line of the buildings that border the basin. . . . Far below, at the water's edge, another continuous line of white lights springs to view, completely surrounding the great water court, mirroring itself. . . . A search light . . . sweeps down upon the huge "Republic." . . . And now the playing of the electric fountains begins, a play of water and of fire.[13]

"Turn your eyes to whatever building you please," wrote another visitor, "you see hosts of suns, moons, and satellites illuminating this model of an earthly heaven."[14]

"Night is the magician of the Fair," wrote John Ingalls in *Cosmopolitan*. "By day the illusion is not complete. The outline and masses, the groups and spaces, the vistas and perspectives . . . are superb and inspiring; but the sun is pitiless and reveals too much."[15] Night and nighttime lighting reduced the fair to human proportions. Ingalls noted too many people wandering around in the day, dazed by the scale of the exposition, overwhelmed in the fatigue brought on by its possibilities. "The monotonous multitudes that incessantly wander to and fro, apparently without interest or enjoyment in the marvels by which they are surrounded, become oppressive." But, in the darkness the fair became "mysteriously luminous." Ingalls continued, "Distant domes grow translucent with interior flame. Cornice and pediment and colonnade are traced in golden beads of fire. The palled pinnacles are etched upon the ebony sky, and, suddenly, 'the long light shakes across the lakes, and the wild cataract leaps in glory.'"[16]

Theodore Dreiser, then a young reporter for the *St. Louis Republic*, accompanied a group of schoolteachers to the fair. Standing at night on the balcony of his hotel, the western sky over the Midway Plaissance seemed "one vast blaze of lurid fire, the siege of Sebastopol." To the east the giant exhibition buildings were in full view, "their many domes and spires lighted up by thousands upon thousands of electric lamps, until the whole appears as some enormous piece of silk lace and drapery, spun of flaming silver and molten gold."[17] For Dreiser, the daylight also made the fair uninteresting. But after sunset "the World's Fair is superb and then alone," he wrote. "Soon its spires are fringed with ropes of fiery gold until the walls are brighter illuminated than by day and the shadows of the trees melt into nothingness. How the statuary gleam, a silvery brightness in the glare of the lights. . . . It is then that the Fair Grounds seem a garden of the gods."[18]

Frederick Law Olmsted, who toured the grounds repeatedly, was unsettled to find by day that crowds "wore a tired, dutiful, 'melancholy air.'"[19] More excitement was needed, he urged Burnham. That excitement could be found on the Midway Plaissance. A centerpiece of the area was the giant Ferris Wheel outlined at night by incandescent bulbs. There also were the Austrian, Egyptian, Irish, German, and Samoan "villages," the Moorish Palace, the replica of Mount Vernon, a Buddhist temple, and restaurants, beer gardens, and music halls galore. As the heroine in Clare Louise Burnham's novel, *Sweet Clover, A Romance of the White City,* describes: "In the Midway it's some dirty and all barbaric. It deafens you with noise; the worst folks in there are avaricious and bad, and the best are just children in their ignorance, and when you're feelin' bewildered with the smells and sounds and sights, always changin' like one o' these kaleidoscopes . . . you pass under a bridge—and all of a sudden you are in a great beautiful silence. The angels on the Womans's Buildin' smile down and bless you."[20]

TRANS-MISSISSIPPI AND INTERNATIONAL EXPOSITION

At Omaha's Trans-Mississippi and International Exposition in 1898, visitors entered the Grand Court through the "Arch of States." There, in the manner of the World's Columbian Exposition, stood another White City with monumental buildings of classical ornamentation, but here they were connected by ivy-covered colonnades. To the south, organized around the Horticulture Building, was an Arcadian park of curved, tree-lined walks and floral displays offering sharp visual contrast to the buildings. To the north was the midway, which featured the "Old Plantation," a village moved from Nashville's Tennessee Centennial Exposition of the previous year. A portion of the village contained "slave cabins" and pretended to show the carefree, happy life of African Americans in the traditional South. Other displays illustrated African American "progress." Nearby, in the home kitchen of the Manufacturing Building, Aunt Jemima served up pancakes.

Only a few foreign exhibits were entered in the fair, since few countries were convinced that a small city like Omaha could organize and finance an exposition. Fair managers, therefore, relied upon such midway attractions as the Chinese Village to provide international flavor.[21] A centerpiece of the midway was the great see-saw, also moved from Nashville, which elevated two viewing platforms some 200 feet into the air. The biggest draw was the Indian Congress, which brought to the fair representatives of various Native American tribes as well as exhibits provided by the Bureau of Indian Affairs. The Apache war chief Geronimo, furloughed from prison to attend, was regarded as a major celebrity. A number of "Wild West" shows were featured at the fair, including that of Buffalo Bill Cody.

Visitors could see "the power of electricity" inside the Manufacturing Building, but outside at night electricity became "poetry," observed Octave Thanet, writing in *Cosmopolitan.* "Then, ten thousand incandescent lights make Court and Plaza and Park and Midway streets like softened day; and the lagoon mirrors palaces penciled in fire . . . while fountains rain a jeweled shower, opals or rubies or sapphires or emeralds or diamonds."[22] "And while we look down the vine-wreathed colonnades and the glittering facades at this flower of civilization," he continued, "almost within earshot the Apaches are yelling and dancing around their fires."[23] With this juxtaposition, the superiority of European Americans, and not just their gentry class, was clearly symbolized in their illumination of the night. The dome of the Government Building was outlined in incandescent bulbs, and at top the Goddess of Liberty rose her "blazing torch" of incandescent light. "The perfect diffusion of the light; the carefully studied heights; the absence of arc lights and crude display effects; the art that accentuates instead of overwhelms the beauties of the architecture—all these have produced a most artistic and complete illumination," wrote one visitor.[24] Lighting designer Luther Stieringer, along with many of the fair's exhibits and its Indian Congress, would move on to exhibit the next year at Buffalo.

PAN-AMERICAN EXPOSITION

The first two Niagara power generators went into service in 1895, and a year later a transmission line opened to Buffalo. The impulse to celebrate the progressive "Niagara Frontier" quickly turned into plans for an exposition to celebrate the technological and cultural advances of the entire Western Hemisphere—the Pan-American Exposition. Electricity, and especially electric light, remained very much the central theme of the "City of Light" created for the fair. Free-Renaissance (what some called Spanish Renaissance) styling predominated, the architecture meant to reflect Latin American tastes. Again, large exhibition halls were arranged around a central space—the Court of Fountains. However, this space was much smaller than at Chicago and at Omaha, and there were smaller plazas grouped immediately nearby. Consequently, the scale of things, although visually impressive—especially at night—did not overwhelm and fatigue visitors. Immediately accessible was the midway or "Pan," off to one side. An "Arcadia" of wooded paths surrounding a small lake stood a short distance to the south.

It was the use of color that set the Buffalo fair apart from its predecessors, color that was not applied randomly but rather designed according to a sophisticated plan. Colors delighted, but they also symbolized. Colors were harmonized across building facades, from facade to facade, and from the buildings of one plaza to another. However, as historian Robert Rydell has emphasized, color was also used to differentiate. The strongest, crudest colors marked the vulgarity of the Pan. Elsewhere, refined tints and shades identified cultural sophistication. On the Ethnology Building, brighter colors marked the building's base, softening into pastels higher up. "The color mosaic presented by the fair told the story of the nation's successful struggle with nature and forecast a future where racial fitness would determine prosperity."[25] It was not the African American or the Native American, but the "full American" who was "the most cosmopolitan man—the sum of all races, the union of all forms of talent and gift," wrote one columnist.[26]

Night lighting was engineered to emphasize color. More than 100,000 electric incandescent filament lamps were used in the Court of Fountains, throwing off a light that was "highly diffused" with "no intense points of

OFFICIAL SOUVENIR
MAILING CARD

ELECTRIC TOWER.

With love & kisses from Father

COPYRIGHT, 1901, BY THE NIAGARA ENVELOPE MANUFACTORY   GIES & CO. LITHO. BUFFALO, N.Y., U.S.A.

FIGURE 7.4. A postcard view of the Electric Tower at the Pan-American Exposition, Buffalo, New York, 1901. Every manner of electric lamp and reflective luminaire seems to have been applied in this lighting extravaganza. Light was variously projected and reflected in celebration of Niagara's newly developed hydroelectric power generation.

brilliance" and "without shadows."[27] The fair's Electric Tower was "dedicated to light, which is, spiritually interpreted, the genius of our age," wrote Julian Hawthorne in *Harper's Monthly* (fig. 7.4).[28] The 348-foot structure was lit from within, outlined in incandescent bulbs, and floodlit from without, creating at night a gleaming visual centerpiece observable from nearly every part of the grounds. It tapered upward by stages. At the base flowed a fountain lit blue-green at night, an analogy to Niagara Falls. Two-thirds of the way up, pavilions sat one upon the other, lit so as to silhouette the pillars. Atop the tower, the female figure of Electricity looked down, arms upheld. In the adjacent Electric Building, telephones connected to Niagara, allowing visitors to hear the roar of the falls at the Cave of the Winds.

On the Pan, the Acetylene Building was lit in a blinding white light, quite in contrast to colored subtleties elsewhere. There, the biggest moneymaker proved to be the "Trip to the Moon." "Yes, verily, there we sit, while the marvelous vessel waves its wings, and far below us, with its electric lights shining, lies the terrestrial city of Buffalo," Hawthorne wrote.[29] Passing through a lightning display with clouds and thunder, the passengers finally reached the moon and disembarked to visit the "Castle of the Man in the Moon," where "Moon Maidens" danced and dwarves cavorted on walls of green cheese. After the fair closed, this show was moved to Coney Island as the centerpiece of the new Luna Park. Unfortunately, the assassination of President William McKinley at the fair in September 1901 dampened festivities, and the fair subsequently closed early.

LOUISIANA PURCHASE EXPOSITION

The 1904 Louisiana Purchase Exposition in St. Louis was gigantic, twice the size of Chicago's White City. Incorporating lessons learned at Buffalo, planners constructed mammoth exhibition halls but grouped them closely together along three axial plazas radiating south and west from a ceremonial "Terrace of the States." The midway, or "Pike," was located immediately south of the exhibition halls. Various classical revival motifs were evident in the fair's architecture. The middle plaza contained the customary "Grand Basin," with the Palaces of Electricity and Education fronting the basin on opposite sides. On the facade of the Electricity Building were loggias and galleries from which spectators could view a large cascade that was brightly illuminated at night. On pedestals over the building's front porticoes were large statues symbolizing Light, Power,

Speed, and Heat. On corner towers were statuary ensembles symbolizing Light and Darkness.[30]

The Terrace of States was lit by more than 200,000 incandescent filament lamps, producing the primary visual effect of silhouetting of the pillars in the colonnades. "Lamps are placed 15 inches apart on the rear side of the columns, and the light striking the building walls from 10 to 20 feet behind, causes the columns to stand out sharply against a brilliant white background."[31] Three sets of lights were used with white, amethyst, and emerald filters so that the color effect could be varied. Searchlights shined from the tops of the Electricity and Education Buildings, and the water of the basin was illuminated from beneath. After unsuccessful experiments with Cooper-Hewitt tube lamps, incandescent filament bulbs with emerald filters were used.[32]

The Pike was the most successful midway yet. As pioneered at Omaha and continued at Buffalo, "ethnological" displays were lavish. The Philippine exhibit was an entire walled town of some forty acres. There was a replica of the Grand Trianon at Versailles, a large Roman villa, and Chinese, Irish, Japanese, and Spanish villages. Entertainments included the Streets of Cairo, Gay Paree, and the Streets of Paris. An Indian Congress was held, and every day "Custer's Last Stand" was simulated. The giant Ferris Wheel was brought down from an amusement park at Chicago, and every night Hale's Fire Fighters put on a fire extravaganza. The hot dog was popularized and the ice cream cone was invented at the St. Louis fair. A song was even written:

> *Meet me in St. Louis,*
> *Meet me at the Fair.*
> *Don't tell me the lights are shining,*
> *Any place but there.*[33]

PANAMA-PACIFIC INTERNATIONAL EXPOSITION

Despite its lengthy title, Portland's 1905 Lewis and Clark Centennial and American Pacific Exposition and Oriental Fair was small. Perhaps its most celebrated feature was the giant electric sign erected on one of the city's hills, which simply displayed "1905." The fair, however, did stimulate what became the annual Portland Rose Festival. Seattle's 1909 Alaska-Yukon-Pacific Exposition (commonly called "A-Y-P"), was also small. A city park was designed by John C. and Frederick Law Olmsted, Jr., into which both temporary and permanent exposition halls were placed. The buildings, which were of classical styling, were organized around a modest central courtyard, and immediately beyond was an Arcadian world of forest and glen. The fair advertised itself as the "World's Most Beautiful Exposition." The "Pay Streak" also was given an integrated architectural styling dubbed "Jap-Alaskan," with strings of Japanese lanterns connecting totem poles along its main axis. However, it would be at the San Francisco Panama-Pacific International Exposition of 1915 that real innovation in lighting would occur. At San Francisco, floodlighting was introduced on a massive scale for the first time.

The 1915 San Francisco fair celebrated the opening of the Panama Canal. It was comprised of large exhibition halls arranged around a multiple of courtyards, as had become standard. Unlike previous fairs, however, there was no single central focus—save, perhaps, the "Tower of Jewels," which stood taller than other towers, nearly every building having an elaborate pinnacle of some sort. At night searchlights provided a sense of integration. Scintillators of Walter D'Arcy Ryan's design operated as they had at the 1909 Hudson-Fulton Celebration in New York City. The San Francisco fair was intended to be a lighting extravaganza, but Ryan, the lighting designer, rejected both the outline lighting of festooned incandescent bulbs and the use of open electric arc lamps. Only in the "Joy Zone" was such garish lighting permitted. Floodlighting was adopted instead, much of it screened, filtered, and reflected in novel ways. High-pressure gas mantle lamps were used for the first time, as were high-wattage incandescent tungsten filament lamps.[34]

At night, light was softly diffused across building facades, its intensity increasing upwards along the numerous towers with color schemes varying from court to court. Hooded luminaries mounted on posts lit the "Court of Abundance." Ryan himself wrote: "Soft radiant energy is everywhere; lights and shadows abound, fire spits from the mouths of serpents into the flaming gas cauldrons and sends its flickering rays over the composite Spanish-Gothic-Oriental grandeur. Mysterious vapors rise from steam-electric cauldrons and also from the beautiful central fountain group symbolizing the Earth in formation."[35] Facade lighting was not meant to exaggerate the scale of the exposition buildings or their connecting spaces. Rather, floodlighting was used to set moods of intimacy. Lighting was not intended to strike awe so much as to seem beautiful. However, the overhead beams of light created by the searchlight scintillators remained awe-inspiring. The colors of the rainbow changed overhead from one hue to another as the beams of light shifted direction in the night sky, creating a simulated aurora borealis. Over 2 billion candlepower was employed.[36]

San Francisco was not the only West Coast city celebrating the opening of the Panama Canal with a fair. The year 1915 also saw the opening of San Diego's Panama-California Exposition. Like Seattle's fair, the San Diego endeavor was very small and, as at Seattle, the grounds were designed to be a permanent park, in this case Balboa Park. Indeed, many of the fair's Spanish revival buildings are still used today to house museums, including the Museum of Man. The "Science of Man" was the exposition's theme, and ethnological and archaeological exhibits predominated. Pseudoscience prevailed on the "Isthmus," however, showcased in its entertainment attractions. Lighting, meanwhile, did not emphasize buildings so much as their surrounding vegetation. Twenty years later the exposition was revived as the California-Pacific International Exposition, using many of the same buildings and, of course, the same park setting. One visitor was impressed by a "canyon full of palms and ferns" turned by lighting into "rare tints and luminous shadows," a formal garden from "the courts of old Spain" with "its brilliant flowers and sparkling fountains," and the patio of a Spanish Renaissance palace with "the delicate fronds of tropical plants and ferns seen in lacelike silhouette against warm amber-lighted walls."[37]

CENTURY OF PROGRESS EXPOSITION

Chicago's 1935 Century of Progress Exposition was the next large fair to use innovative lighting, Philadelphia's 1925 Sesquicentennial Exposition having proved a weak attempt to relive the glories of 1876. Among the light displays at Philadelphia, however, was a giant replica of the Liberty Bell covered by some 26,000 electric incandescent bulbs. Part of Philadelphia's problem had stemmed from fair planners' reliance on the old formula of large exhibition halls traditionally ornamented and organized around formal courtyards. Chicago would not repeat that mistake. The Century of Progress Exposition would be radically new in terms of architecture, layout, and lighting. Its architecture was streamline modern, its layout almost totally rejected formal geometrics, and its lighting system made extensive use of the new electric gaseous discharge lamps, especially neon tubing.

The use of "modern" architecture at the fair capitalized on new materials—asbestos and gypsum board as well as plywood—prefabricated and attached to steel frames. Buildings were produced with smooth wall surfaces, brightly colored and largely devoid of windows and other openings, excellent surfaces off of which to reflect light. Thus lighting was far from an afterthought. Much of the fair was conceptualized from the beginning as a lighting display. Its nighttime opening signified as much. A ray of light from the star Arcturus, which had originated forty years earlier at the time of the World's Columbian Exposition, was captured by a telescope and converted, by a photo-electric cell, into an electrical impulse used to ignite a giant fireworks display.[38]

The grounds crowded informally along the Lake Michigan shore. One long strip of exhibits stretched for over two miles, not unlike the new commercial strips then emerging along the nation's highways. Across a lagoon at the north end, on a peninsula running parallel to the shore and linked to it by three causeways, was another, much shorter strip. The biggest exhibit halls were built by corporate sponsors (General Motors, Goodyear Tire and Rubber, Firestone), by foreign governments (China, Czechoslovakia, Italy), and by trade associations (the electrical and gas industries, for example). Buildings were shaped to reflect their sponsors, using architecture as sign. Joseph Urban, the architect and stage designer whose job it was to color-coordinate the fair, chose twenty-four hues—green, two blue-greens, six blues, two yellows, three reds, four oranges, two grays, white, black, silver, and gold.[39] Bold splashes of color suggested a carnival atmosphere, not just for the designated midway (which existed primarily as an amusement park with rides) but for the entire exposition.

Walter D'Arcy Ryan coordinated the fair's lighting, the purpose of which was to (1) enable visitors to familiarize themselves with and negotiate the layout of the grounds, (2) endear them to the new style of architecture, and (3) extenuate the color pallet. In Ryan's own words, the bold plan involved "large buildings constructed of new materials, windowless, interconnected with bridges, and with viewing terraces at many levels," "dissymmmetry" of both building form and grouping, and "long viewing distances."[40] Light fixtures of modern design were placed in full view, and there was no attempt to conceal or screen their structural elements. As might be expected, Ryan's scintillators also sent searchlight beams radiating across the sky.

The fair's central feature was the Sky Ride. Two giant sixty-story towers were connected at the twenty-fourth story by cables from which "rocket cars" were suspended for trips across the lagoon (fig. 7.5). Visitors could also continue to the top of each tower, where there were observation platforms. Whereas the giant Ferris wheel at the World's Columbian Exposition and the great see-saw at the Trans-Mississippi Exposition had been relegated to their respective midways, here the Sky Ride attraction was at the very heart of the fair. The exposition's official guide described the view down: "If you stand in one of these observation rooms at night and look down, you gaze upon a magic city that seems to float in a vast pool of light. From the towers, great searchlights sweep the sky, the lake, and over the great city to the west, to clash with other massive beams of light."[41] At the Pan-American Exposition, the "Trip to the Moon" had involved glimpses down on a model of the City of Buffalo, but here the actual city appeared to be a model at the visitor's feet.

Over 75,000 feet of neon tubing was installed at the fair, with over a mile and a half in the Electric Building alone.[42] The west facade of that structure contained 4,600 feet of blue neon in seven vertical columns each fifty-five feet high. At the building's grand staircase, four setbacks were illuminated indirectly by 2,100 feet of red tubing, spreading a rich glow of color. The 176-foot pylon tower in the Hall of Science was lit in deep blue on the east and north and in bright red on the south and west, using tubing concealed behind wooden appliqué. The Hall of Social Science was lit in silver and blue. More than 15,000 incandescent lamps were used, in-

FIGURE 7.5. Sky Ride at the Century of Progress Exposition, Chicago, 1933. This postcard was published well before the fair opened, as an idealized anticipation of what was to come. It demonstrates the importance assigned by fair planners to nighttime lighting.

stalled on posts to light the walks and roadways and in the giant searchlights.[43] The world's largest electric incandescent filament lamp—50 kilowatts—was placed on display.

The Texas Centennial Exposition at Dallas and the Great Lakes Exposition at Cleveland, both held in 1936, copied the 1933 Chicago fair in both architectural design and in lighting. The "Esplanade of States" at Dallas was lined with modernistic fifty-foot pylons hung with banners. Nearly fifty miles of luminous tubing was used overall, illuminating the grounds and buildings in shades of red, yellow, and green.[44] The major oil companies—Conoco, Gulf, Magnolia, Sinclair, and Texaco—along with Chrysler, Ford, and General Motors, built large pavilions. In Cleveland, the "Court of the Presidents" along Lake Erie's shore was lined by sixty-foot pylons topped by eagles. Elsewhere pennants hung on tall masts like those of ships and lined walks and roadways. Floodlighting, rather than neon, dominated, along with mercury vapor lamps, widely used for the first time at a fair.[45] Double-silhouette signs were set against the flat facades of buildings, in keeping with the use of modern architecture with a nautical flair. Again, searchlights at both Dallas and Cleveland sent beams of colored light overhead.

GOLDEN GATE INTERNATIONAL EXPOSITION AND THE WORLD OF TOMORROW

The two fairs of 1939—San Francisco's Golden Gate International Exposition and New York City's World of Tomorrow—had very different personalities. At San Francisco, streamline modern was combined with historical design motifs. Planners sought a novel blend of occidental and oriental decor suggestive of the varied cultures ringing the Pacific Basin. Cambodian, Burmese, Incan, and Mayan ideas were most recognizable. Color was applied to plastered wall surfaces, the plaster given exaggerated luminosity through the use of vermiculite, which was essentially chipped mica. Fluorescent tubular lamps, never used at a fair before, were used to bring the whole complex alive at night; the most spectacular effects were achieved using "black light." The exposition was located on Treasure Island, reached by the San Francisco–Oakland Bay Bridge. Acording to one journalist, "a luminous dream city that appears to float on the black velvet of San Francisco Bay" was created. "White and gold and pastel shades of shimmering light reflect on waves, while tall fingers of colored brilliance comb the night skies."[46] The play of color on buildings emphasized off-whites, amber, lemon yellow, green, magenta, apricot, peach, pink, gold, blue, and red, each court with two or three harmonious colors combined. Tints and shades rather than primary colors were used throughout. Some 2,300 tubular fluorescent lamps, most in shades of green, blue, pink, and yellow, were employed, mounted with cylindrical parabolic reflectors.[47] Mercury vapor floodlamps were also used.

New York City's 1939 World of Tomorrow, as the name indicates, was futuristic. The exposition sprawled over an extensive site adjacent to, but turned away from, Long Island Sound. Buildings, little more than plain geometric boxes, stood isolated in greenswards—they stood out as objects set boldly in space, the basis for much of America's evolving modern automobile-oriented suburbia. It was a return to a gigantic scale, overwhelming to the weary pedestrian. The linear axes of a giant grid of streets brought a basic logic to the plan, and a sense of center was achieved with the giant "Trylon" and "Perisphere," which loomed over the fair's other buildings. The Trylon, a slim 700-foot triangular tower, rose above the Perisphere, touted as the biggest ball ever built. Inside the Perisphere, a moving platform took spectators over a miniature "Democracity," described as "the City of Tomorrow Morning."[48]

The World of Tomorrow was a fair of modeled urban landscapes. In the Consolidated Edison Pavilion, visitors could see the world's largest diorama, about one city block in length and three stories tall. Designed by Walter Teague and called "The City of Light," it portrayed New York City at night.[49] The big attraction of the fair was General Motors' "Futurama." Visitors sat in moving chairs that were swept up, over, and around another gigantic model—the "World in 1960." Its tall skyscrapers, located in greenswards, were connected by superhighways, on which automobiles and trucks traveled.[50] It was as if such models were necessary not just to communicate about the present and the future but to return the fair itself to human scale.

Planners sought to make the fair's buildings appear to be "light sources themselves," rather than mere reflectors targeted by light. Yet targeting with fluorescent light is exactly the method that was employed. Nearly one-third of the lighting was fluorescent. For example, the Metals Building was sheathed in light from blue-green fluorescent lamps that were screened from view. The Trylon and Perisphere, on the other hand, were lit by a battery of 200 mercury vapor floodlamps mounted on the roofs of nearby buildings. Although the exposition's midway was called the "Great White Way," named after the city's brilliantly lit Broadway at Times Square, lighting was not the primary attraction there. Thrill rides such as the giant 250-foot parachute jump captured the public imagination instead.

THE LATER FAIRS

The last large American fairs were held in the 1960s. Seattle's Century 21 Exposition, in 1962, was again futuristic in its theme. Its central feature was the Space Needle, from which viewers could look down on a real city by day or night (as at Chicago in 1933). A monorail connected the tower with the nearby downtown district. Mercury vapor floodlights were mounted on the space needle itself, and the revolving top, containing a restaurant, was outlined by fluorescent tubing. In the eleven-story Washington State Colosseum, the "Bubbleator" was created—a large plastic sphere into which visitors were lifted by hydraulic elevator. Surrounding the dome was a huge, free-form sculpture comprising thousands of aluminum cubes across which light moved in changing colors and rhythms. Spectators were then led down a series of winding pathways from which they could glimpse another model—"Seattle in the Year 2001." Artificial Light sequenced from dawn into day and then through sunset into night again.[51]

The New York World's Fair of 1964 was very much like its 1939 predecessor, being located on the same site and overseen once again by Robert Moses, builder of many of the city's bridges, expressways, and public housing complexes. The fair's central feature was the Unisphere, a permanent structure donated by the United States Steel Corporation. The giant world globe rose up and out of a large reflecting pool surrounded by a double ring of water jets, all brightly lit at night. Careful viewing of the globe revealed light and darkness alternately sweeping across the continents, recreating the diurnal cycle. A nightly festival combined music, water, electric light, and fireworks. Mercury vapor lamps dominated the lighting of the grounds, giving off an eerie blue-green light. Backlit signs and backlit wall partitions were widespread—these were made from acrylic plastics, the kinds of materials and lighting effects already common in newer office buildings and shopping centers nationwide. Indeed, the visitor to the fair saw few unfamiliar lighting effects; the festive city and the actual city were substantially merged as light display.

TRADITIONALLY, AMERICANS' FIRST EXPOSURE to new lighting technologies, and their extensive application at the landscape scale, was usually at a fair. Not only were new lighting technologies displayed at exhibits in the fairs, but, more importantly, they were also extensively used in lighting the buildings and the grounds. Americans went to the world's fairs to gain an appreciation for the past and to anticipate the future. There they encountered demonstrations of what America pretended to be and what America intended to become. Yet, in reality, Americans did not reconfigure their cities in the image of Chicago's "White City," except for relatively few civic centers of the "City Beautiful" ilk. Nor did American cities come to be ornamented like the exotic courtyards of the San Diego or San Francisco expositions. But, to some degree, Americans came to illuminate their houses, offices, and stores in ways reminiscent of the fairs. The spectacular effects of searchlights, floodlights, and beacons, for example, were later seen at night in most American cities if only in token display.

Artificial illumination became essential to the design of the fairs. From 1893 onward, fairs tended to be epitomized in the public mind by night scenes. Exposition planners embraced light, intending their fairs to be light shows during the evening hours. Electric light could dazzle crowds, sending forth clear social messages. Here modern technology—complex, sophisticated, and somewhat mysterious—was made to symbolize national superiority. Every visitor to every fair was invited to join in the celebration and to partially redefine himself or herself in the experience.

As light was used to captivate the multitudes, so it was also used to win their support for future-oriented projects. Nowhere but at the fairs was the power of nighttime lighting used so forcefully for this purpose. Before he was struck down by an assassin's bullet at the Pan-American Exposition in 1901, President William McKinley had emphasized the connection between America's great fairs and the nation's sense of progress. "Expositions are the timekeepers of progress," he said. "They record the world's advancement. They stimulate the energy, enterprise, and intellect of the people and quicken human genius. They go into the home. They broaden and brighten the daily life of the people."[52]

Many lighting schemes demonstrated at the world's fairs found ready application in street lighting. However, innovative architectural lighting proved less influential. The subtleties of diffused lighting emphasizing subdued pastel hues, as characterized many of the later fairs, proved inappropriate to streets that required bright illumination for automobiles. The expositions demonstrated varied place-making possibilities for night light, but the rise of the automobile limited the nation's lighting choices. However, the lighting schemes of the world's fairs stand as reminders of what might have been. They are also suggestive of what yet could be.

CHAPTER EIGHT

Night at the Amusement Parks

Large expositions closed, but many of their attractions did not. They merely moved on to appear at other fairs or found permanent homes in the nation's new amusement parks. From 1890 through World War II, nearly every large city in the United States had its brightly lit nighttime "fun park," and most of the larger cities were home to several competing facilities. Many parks were operated by or in affiliation with the traction companies—some, located at the ends of streetcar lines, were called "trolley parks." In turn, many traction companies were owned by public utility companies, and to them the parks represented another means of promoting electrical power. The nation's largest amusement parks included Atlanta's Ponce de Leon Park, Boston's Revere Beach, Chicago's Riverview and White City, Cleveland's Euclid Beach, Denver's Manhattan Beach, Philadelphia's Willow Grove, and St. Louis's Forest Park Highlands.

Amusement parks replicated at a larger scale activities long associated with beer gardens, dance halls, and music halls—in other words, the parks revolved around eating, drinking, dancing, and vaudeville. The carnival-like boardwalks at public beaches also served as prototype. Indeed, many beach resorts had amusement parks that flourished as protected compounds where the middle classes could enjoy the fun without having to mix with the rougher elements attracted to boardwalk amusements. Admission charges at these amusement parks kept out the hoi polloi. The midway attractions were borrowed from the world's fairs; especially popular were the rides that rapidly lifted, tilted, swirled, or plunged visitors.

Some resort cities evolved clusters of such amusement venues, for example Atlantic City in New Jersey and Long Beach in California, but New York City's Coney Island had the largest concentration. Coney Island set the entertainment tone, perfecting the hallmark gravity-defying rides and other innovated techniques of promotion which would last for half a century or more and define the amusement park primarily as a working man's resort. In the 1950s, Walt Disney launched an era of theme parks oriented to more affluent audiences but built around many of the same entertainments that sustained Coney Island. Recently, Disney's efforts have culminated in the Epcot Center, a permanent "Worlds's Fair" near Orlando. There, as at Disneyland and Walt Disney World, Disney accomplished what the earlier amusement parks did not. Walt Disney had linked the amusement park idea with modern entertainment media, especially film and television—and in so doing he tapped the energy of an industry that might otherwise have been his amusement park's biggest competitor. It could be argued that what had undermined the traditional amusement park in the years before World War II was the completion of the nation's motion picture theaters.

ILLICIT ROOTS

In the nineteenth century, gas lighting on many city streets became synonymous with the illicit. Lit streets encouraged retailers and other merchants to extend their hours, but lighting's most immediate impact was not necessarily on "legitimate" business so much as on the "illegitimate," especially the entertainments deemed vulgar by society's moral and cultural elites. Lit streets were seen as encouraging the expansion of city districts where the traditional after-dark entertainments of drinking, eating, and attending the theater were coupled with prostitution, gambling, and other vices. In New York City this type of neighborhood was epitomized by the Bowery, originally the commercial center of New York's large German community. The Bowery was home to numerous beer gardens and music halls that were originally frequented, in the German tradition, by entire families, especially on Sunday afternoons.

As affluent second- and third-generation German American families moved northward with the expansion of the city, they were replaced by less affluent newcomers and Bowery Avenue quickly evolved into a working-class entertainment zone with largely male-oriented attractions. Some affluent adventurers did continue to frequent the place, however, enticed by what were, for the period, shocking revelations published in such books as George G. Foster's *New York By Gas-Light*.[1] Foster's task was, in his words: "To penetrate beneath the thick veil of night and lay bare the fearful mysteries of darkness in the metropolis—the festivities of prostitution, the orgies of pauperism, the haunts of theft and murder, the scenes of drunkenness and beastly debauch, and all the sad realities that go to make up the lower stratum—the underground story—of life in New York!"[2] New York was not alone. Every large city had its illicit entertainment zone: there was the Barbary Coast in San Francisco, the original "skid row" in Seattle, New Orleans's Rampart Street, and Chicago's "Levee."[3]

In the Bowery, after darkness had settled in, "broad rays of light stream brightly into the open air from the stores, restaurants, and saloons. . . . only the smaller retail shops, the drug stores, the saloons, restaurants, and tobacconists remain open, but these are numerous enough to give a brilliant coloring to the street with their bright lights and elaborately-decorated windows," wrote one observer.[4] Here was the after-dark world of the male *flâneur* and the streetwalker, the female object of the predatory male gaze. "Loud-voiced and foul-tongued" women did not "hesitate to accost men, and too often succeed in inducing them to accompany them to one of the dance-houses, or 'gardens,' which abound in the side streets, and in whose pay these women are."[5] There "girls lounge about in the midst of the smoke; do not hesitate to sit on the laps of gentlemen, and are always ready for one of the foaming glasses of beer."[6] In such writing, the origins of temperance and prohibition are evident.

For many middle- and upper-class visitors to the Bowery, the place exuded more visual spectacle than vice, especially after the building of the "el," which allowed trains to constantly move overhead. Electricity enhanced the magic of the street. "The scene at night is indescribable," wrote William Archer, a visitor from England. "The air seems supersaturated with electricity, flashing and crackling on every hand. One has a sense of having strayed unwittingly into the midst of a miniature planetary system in full swing, with the boom of the trolleys, in their mazy courses, to represent the music of the spheres."[7]

CONEY ISLAND

Coney Island (named for the "conies," or rabbits, which dominated its dunes) had developed into a diverse resort by 1880. On the eastern end, the upscale "Manhattan Beach" and "Oriental Beach" resorts were created around large hotels with landscaped lawns, broad verandas, and opulent dining and sitting rooms. There the gentry came to walk the beach, watch the sunsets, and eat the celebrated oyster dinners. To the west was "Brighton Beach," catering more to the middle classes with a large hotel, several public bath houses, dance halls, vaudeville theaters, and circus attractions. Farthest to the west was Norton's Point, early a haunt of gamblers and prostitutes. Here males of various classes sought the same illicit entertainments that the Bowery offered.[8]

Created along the beach "on the Bowery" (named after the Manhattan street) was an amusement zone much like Philadelphia's "Dinkeyville." The street's name was later changed to Surf Avenue. With the closing of the Centennial Exposition, many of the Philadelphia fair's attractions were moved to Coney Island, prime amongst them the 300-foot observation tower—which would remain the tallest structure in New York City for nearly a decade. A hotel and shopping complex designed as a giant elephant and a duplicate of Chicago's Ferris wheel were other prominent attractions established by 1895. Gradually the whole of Coney Island evolved as a kind of "honky-tonk" district, peopled on a daily basis through railroad and transit connections to the centers of Brooklyn and Manhattan.

The amusement park idea was introduced to Coney Island in 1897, when George Tilyou built Steeplechase Park. In order to exclude the rowdier elements and encourage family patronage, Tilyou enclosed his grounds and charged admission. He provided a clean, landscaped park with amusement rides, side shows, music and dance halls, and restaurants, all coordinated with a unified architecture emphasizing fantasy and exotica. Hints of controlled sensuality combined with respectable family entertainment proved to be a magic mix. The centerpiece of the park was the mechanical horse race, in which enthusiasts could sit upon wheeled mounts and, by shifting their weight, control speed in actual races down a dozen or so tracks.

FIGURE 8.1. Entrance to Luna Park at Coney Island in New York City, circa 1910. "Surf Avenue," claims the postcard's caption, "is illuminated with 1,000,000 electric lights and is the most brilliantly lighted thoroughfare in the world."

In 1903, Frederic Thompson and Elmer Dundy, competitors at Omaha's Trans-Mississippi Exposition but partners at Buffalo's Pan-American Exposition, brought their "Trip to the Moon" attraction to Coney Island and installed it in their new Luna Park (the park was actually named for Dundy's sister, but people soon assumed it had been named for moonlight and artificial illumination). Luna Park's lighting included 250,000 incandescent filament lamps festooned across the facades of various buildings—the Eskimo Village, the "Shoot-the-chutes," the "Trip to the North Pole," the German Village, the Canal of Venice, and the Grand Ballroom. The whole park was a gigantic stage set with the architecture of the absurd, all brought to tantalizing brilliance through a sophisticated application of electric light, an attraction in itself. At the center of the park was still another giant Electric Tower.

Dreamland Park, built next to Luna Park in 1904 by William H. Reynolds, used a million incandescent bulbs, 100,000 of which were on its own Electric Tower. The structure could be seen at sea for over thirty miles, and its searchlights had to be dimmed because they interfered with ship navigation. Dreamland's attractions included "Lilliputia" (a complete town scaled to fit several hundred dwarf actors), the "Infant Incubator" (with its premature babies), the "Fall of Pompeii," the "Frank Bostock Wild Animal Show," and the "Leap Frog Railroad." Each night the "Baltimore Fire" was acted out with flames and pyrotechnics.

At Coney Island, electric lighting was first promoted for "electric bathing," extending use of the beach into the nighttime hours through the use of arc lights. With the opening of the amusement parks, however, spectacular lighting quickly stole the nighttime scene. Travel writer James Huneker was very much impressed by the gaudy lights. Coney Island, with its "vulgarity, its babble and tumult, is a glorified city in flame," he wrote.[9] "The view of Luna Park . . . suggests a cemetery of fire, and mortuary shafts of flame. At Dreamland the little lighthouse is a scarlet incandescence. The big building stands a dazzling apparition for men on ships and steamers out to sea. Everything is fretted with fire. Fire delicately etches some fairy structure; fire outlines an Oriental gateway; fire runs like a musical scale through the octaves, the darkness crowding it, the mist blurring it."[10] Huneker stayed through the night, along with a Saturday crowd in late summer numbering in the thousands, sleeping on the beach after the lights turned off. "A muggy moon shown intermittently over us, its bleached rays painting in ghastly tone, the upturned faces of the sleepers."[11]

Meanwhile Maxim Gorky, visiting New York City, was hostile to what he saw at Coney Island. "With the advent of the night a fantastic city all of fire suddenly rises from the ocean into the sky." he wrote. "Thousands of ruddy sparks glimmer in the darkness, limning in fire, sensitive outline on the black background of the sky, shapely towers of miraculous castles, palaces, and temples. Golden Gossamer threads tremble in the air."[12] Nonetheless, Gorky categorized it all as a kind of opium for the masses. "The visitor is stunned; his consciousness is withered by the intense glow," he continued. "People wander about in the flashing, blinding, fire intoxicated and devoid of will. A dull-white mist penetrates their brains, greedy expectation envelopes their souls."[13] One needed to struggle to retain a sense of individuality; the light revealed not delight but a dismal reality of exploitation. "The city, magic and fantastic from afar, now appears an absurd jumble of straight lines of wood, a cheap, hastily constructed toy-house for the amusement of children."[14]

Reformers associated bright lights and leisure-oriented mechanical devices with moral depravity.[15] They would have had Coney Island turned into a park, following the examples of Central Park in Manhattan and Prospect Park in Brooklyn. The gentry preferred to use leisure time in order to cultivate lofty and refined desires. Coney Island, like the world's fair midways, attracted the lower classes, people who did not appreciate the genteel, highly introspective celebration of nature kept in some pristine state or the refinements of classical architecture. These masses were easily attracted to the fully urban, fast-paced, extroverted enjoyment of the sensual and the innovative. At Coney Island, as at the world's fairs, they could step out of everyday routines under dominant cultural authorities and relax in new, less-regulated social contexts.

Against the gentrified values of thrift, sobriety, industry, and ambition, Coney Island encouraged, if only temporarily, the release of extravagance, gaiety, and revelry.[16] What is more, in this stage-set world, patrons were invited to play as actors rather than watch as mere spectators. It was, as Gorky argued, thoroughly escapist, a pressure valve for the masses. Social tensions were relieved rather than channeled toward conflict and social change. A system was functioning, it could be argued, in which capitalists exploited the masses' desire to release their frustrations.

WALT DISNEY

Coney Island reached its high point around 1910. Fire destroyed Dreamland the next year, and it was not rebuilt. Surf Avenue dissolved into shabby remnants over ensuing decades, many of which were converted to marginal nickel and dime amusements. Walt Disney, appalled by what Coney Island and most of the nation's amusement parks had become, introduced a new kind of highly sanitized "theme park" built around historical and futuristic ideas. Though novel to a new generation of affluent Americans, Disney's venues were, nonetheless, highly reminiscent of the earlier amusement parks and past world's fairs.[17] Disneyland, in Anaheim, California, near Los Angeles, and the later Walt Disney

World, near Orlando, were the pioneer parks. They were joined quickly by a host of imitators: Chicago's Great America (with sister parks at Washington, D.C., and San Francisco), Cincinnati's King's Island, Dallas's Six Flags over Texas (with sister parks at Atlanta, St. Louis, and Houston), Nashville's Opryland, Cedar Island, between Cleveland and Toledo, and Freedomland near New York City.

Opened in 1955, Disneyland was promoted through Walt Disney's weekly television show of the same name, and the ABC television network held a one-third ownership in the park. Both the show and the park were segmented along four themes: Adventureland, Frontierland, Fantasyland, and Tomorrowland. Like many world's fair venues, Adventureland was a celebration of exotica, where visitors were able to experience a tropical rain forest from within a giant tree-house. Frontierland capitalized on historical themes in ways reminiscent of the American Indian and "Wild West" shows at Omaha's Trans-Mississippi Exposition. Fantasyland, which included Sleeping Beauty's Castle, reflected Disney's own successes in film animation, emphasizing Mickey Mouse, Donald Duck, and other Disney characters. Tomorrowland, emphasizing futuristic technology, was very clearly influenced by New York City's World of Tomorrow Exposition.[18]

Carefully considered by Disney but ultimately rejected was Lilliputian Land, the town of dwarves that was featured in several world's fairs and a highlight of Coney Island's Dreamland. Also rejected was a recreational zone where large groups could picnic and where, during certain seasons such as at Christmastime, special light displays could be staged.[19] A nightly event at Disneyland, however, was the evening electric light show, in which the park's old-fashioned "Main Street" was decoratively lit by light festoons and Space Mountain, the park's visual centerpiece, was brightly floodlit and framed by exploding fireworks. A trip to Disneyland was not complete without viewing the evening light show, which, in its own right, became one of the park's leading attractions.

Walt Disney World Resort was constructed in 1971 at a much larger scale, to embrace not just a theme park but a set of hotels, two golf courses, several campgrounds, and other resort facilities. It was intended to be a self-contained "total destination resort." The whole was tied together by a monorail, as Seattle's Century 21 Exposition had been. The park's attractions were grouped according to themes, like those at Disneyland, but with more and larger displays. As at Disneyland, there was an obsessive concern with order and cleanliness, and a system of pneumatic tubes was built to evacuate litter and other trash quickly. Walt Disney World's Main Street was scaled down to appeal more to young children and be more convenient for pedestrians. Each evening a parade of light-studded floats was held there, which was coordinated with the fireworks staged over the park's lagoon. Disney had once again combined three of its founder's lifetime interests: his boyhood experiences on the main street of a rural Missouri town, his cartoon and feature film characterizations, and his view of the future.[20]

Opened in 1982, Epcot Center (later shortened to Epcot) was openly cast in the mold of America's world's fair tradition. It was planned as two distinct areas: "Future World," where the "technology of tomorrow" is demonstrated, and the "World Showcase," comprising eight international pavilions representing Britain, Canada, China, France, Italy, Japan, Mexico, and Germany, in addition to an American exhibit. At each pavilion, visitors could sample that particular country's culture. Each national display was given its own lighting treatment—the "warm glow" of the British pavilion contrasting with the "bright vitality" of the French and the "slightly mysterious and serene quality" of the Japanese pavilions.[21] Epcot was not just a destination resort but something of an alternative model for the American city, built around recreation. Indeed, *Epcot* stands for "Experimental Prototype Community of Tomorrow."[22] Theme hotels, a yacht and beach club, and additional golf courses complemented, even dominated, the older Walt Disney World Magic Kingdom nearby.

Each night the day's activities at Epcot are brought to climax through a carefully orchestrated light spectacular. Projectors play animated images on the walls of buildings. Buildings are outlined with strings of Silvestri bulbs. Strobe slights pulse rhythmically to music amplified throughout the park. Fountains display vivid colors that change in phase. Overhead, beams of light are projected up amid fireworks launched from floats at the center of the park's large lagoon, the whole reflected in the water. "The purpose of the late evening show is to say 'Good Night' to the park's guests, to enable them to leave the park with a lovely experience fresh in their minds," wrote one observer.[23] Thus Disney not only developed formulas replicating the world's fairs and the early-twentieth-century amusement parks but also recreated the public light displays of urban festivals.

Amusement parks and their successor theme parks proved enticing to Americans for a host of reasons. As historian Russel B. Nye outlined, they provided alternative worlds into which people could step, temporarily departing from everyday routines.[24] They engendered fantasy in staged settings, where visitors helped act out ongoing dramas. They provided spectacle, encouraging a sense of belonging derived from witnessing just as much as from doing. Amusement parks offered relaxation, further enabling visitors to transcend normal inhibitions. They invited bonding between participants in families and other groups. Their mechanical entertainments carried a sense of "riskless risk," offering the "vicarious terror" of the "first hand hair's-breadth escape."[25] The rise of roller coasters and other thrill rides, with names like "Blue Racer," "Cyclone," "Speed Demon," and "Wildcat," illustrates the public's infatuation with such simulated danger. Finally, the amusement park offered a "total play experience," in which all forms of play (those based on competition, chance, mimicry, or vertigo) were encouraged in a controlled setting.[26]

ARTIFICIAL ILLUMINATION WAS USED TO HEIGHTEN AN AURA of fantasy, the traditional commodity amusement parks purveyed. The exuberant, the sensual, the daring, the irreverent—all were communicated with carefully engineered light. Nothing contrasted more with work routines of the day than the gaily lit entertainments of the night. Nothing seemed to free the individual from the constraints of society like play at night. Nothing elicited the sense of carnival more than vivid, bright color. In the glare of bright lights, people and things were dramatized, providing an open invitation to pleasure. Amusement parks sold "intense, frenetic, physical activities without imaginative demands," wrote one of Coney Island's historians.[27] Bright light made the showy acceptable, the gaudy desirable, and even the ordinary glamorous. Thus for many Americans, it was the display of bright night lights that proved the biggest amusement park attraction. Amusement parks were specialized entertainment zones that glowed and glittered in the night.

The technical sophistication of lighting helped legitimize the leisure-time indulgences of the lower classes as packaged in the amusement parks.

Elite society could not claim technical superiority for their recreational preoccupations, only superiority of acquired taste. Rather than merely imitating the gentry's activities, amusement parks proved that fulfillment could be had in recreations quite apart from elite sensitivities. Indeed, ordinary people felt they could discover and indulge their passions without implicit messages of social inferiority. If urban street festivals promoted a sense of parochial belonging and the world's fairs encouraged a sense of national identity, then the amusement parks gave many groups of people—the less affluent, immigrants arrived from abroad, and rural Americans only recently come to the city—a sense of being part of a newly emergent tradition. That new tradition is what we subsequently have come to call popular culture.

Amusement parks thrived on patrons arriving by train or streetcar. The parks represented visually intense pedestrian worlds for after-work and weekend leisure activities in warm weather, much of the shared visual intensity generated by extravagant nighttime lighting. But automobile ownership soon diverted patronage. Indeed, the automobile itself became an instrument of leisure, as families engaged in recreational motoring to a wide variety of destinations. Working Americans were no longer tied to resorts served by railroads and transit lines. However, many of the visual attributes of the amusement park ultimately came to characterize many of the new roadsides created for and by the automobile. The purpose of these new environments was to engage the motorist, rather than the pedestrian, not only in leisure travel but also in other kinds of movement. I speak, of course, of the exaggerated lighting that came to dominate highway commercial strips. In some ways, the commercial strip replicated the amusement park and world's fair midways—it contained arrays of unobtrusive buildings with exaggerated facades and brightly lit signs, all clamoring for attention at night.

CHAPTER NINE

Landmarks Floodlit in the Night

The term *floodlighting* brings to mind the uniform illumination of building facades and other surfaces. To lighting engineers, however, floodlighting encompasses all manner of light thrown by projectors, including floodlights, spotlights, and searchlights. Closely related to these forms is projector use in beacons. In architectural lighting, floodlights are used at night to emphasize the distinctive elements that lend character to buildings—towers, setbacks, recessed porticos, and so on. Floodlighting is used to illuminate outdoor sculpture, fountains, and other landmarks, allowing these features to define a place even after dark. Athletic events are held as floodlit spectacles in stadiums. Of course, outdoor floodlighting has more than mere decorative or recreational implication. It extends the hours of work, as at construction sites; it extends the hours of service, as at airports. Beacons, for their part, serve to warn, to inform, and to symbolize in a diversity of ways.

As the garish lighting of the midway survived in the nation's amusement parks, so also did the more subdued floodlighting of the world's fairs persist, especially in nighttime architectural illumination. Early floodlighting using electric arc lamps was first demonstrated at the world's fairs. At Chicago's 1893 World's Columbian Exposition, the buildings of the White City were outlined in incandescent lamps and spotlights were trained on fountains and statuary, while searchlights sent beams of light from one building facade to another—the whole choreographed by Luther Stieringer. At Omaha's Trans-Mississippi International Exposition in 1898, Stieringer began, according to one observer, "to show things, not lights, to do not lighting, but light-painting. . . . He hid his lights and threw their radiance on the buildings, bringing out their architectural ornament in bold relief, or painting them luminously in broad washes of white and black."[1] But electric arc lamps were difficult to control with reflectors and prismatic refractors, as were gas mantle lamps. Applications were limited. This dramatically changed, of course, with introduction of coiled-tungsten filament incandescent bulbs.

Incandescent lamps were first adopted on a massive scale by Walter D'Arcy Ryan in his lighting of San Francisco's 1915 Panama-Pacific International Exposition. By concealing the source of light and bringing building facades into brilliant relief against the surrounding darkness, accentuating details of construction, Ryan achieved a new kind of beauty. At night floodlighting could give a building a "real rather than an artificial appearance."[2] Buildings could be made to shimmer in a painterly manner, reminiscent of impressionistic art. Floodlighting was cheaper to install and maintain than earlier "outline" lighting with festoons of incandescent filament bulbs, and it required less electricity to operate. Ryan and other planners appreciated that night lighting should strive to enhance architectural qualities of unity, coherence, balance, scale, and texture. The idea was not to replicate daylight viewing but to make architecture at night equally if not more expressive. Successful compositions balanced light and dark masses through judicious use of framing with accents.[3]

Projecting luminaires were rapidly developed based on the geometric principles of the parabola, a curve of such a shape that light rays from a lamp located at its focus would be reflected outward in lines parallel to its axis.[4] A 500-watt Mazda C lamp located at the focal point of a parabolic reflector could project a narrow beam hundreds of feet in sharp focus. To spread light on nearby surfaces, a lamp had only to be moved in the luminaire to a point behind the focal point of the curve.[5] Once a lamp was diffused in this way, the dark spot at the center of an illuminated surface, where less than full parabolic focus was employed, could be eliminated by using stippled or prismatic lenses. Most parabolic reflectors were of mirrored glass, chromium-plated metal, or polished aluminum. In 1938, parabolic aluminized reflectors sealed behind glass lenses, similar to sealed-beam headlights for automobiles, brought a much more durable and longer-lived floodlamp technology to the market.

Most architects designed buildings to be seen in daylight, usually conceptualizing sunlight as projecting down at a 45-degree angle. Shadows cast by cornices, pillars, and other architectural features were anticipated in such light. Highly concentrated night light, especially when projected from different angles, could distort building facades designed strictly for daylight viewing. If banked in clusters, floodlights could make cornice shadows slope and buildings appear out of plumb. A delicate entasis of classical columns might cast shadows spaced progressively farther apart, disturbing the even march of architectural accent points.[6] If floodlights were separated in an effort to simulate parallel light, multiple shadows could appear. Angles of projection between 6 and 30 degrees above the horizontal were required, otherwise shadows were unduly stretched irrespective of lamp distribution.

LANDMARKS

Edith Wharton, on arriving in Paris in 1912, was discouraged to find the French floodlighting their buildings, a practice she associated with coarse American commercialism. At night, landmarks were "torn from their mystery by the vulgar intrusion of floodlighting," she wrote.[7] Highlighting landmarks was the first permanent use of floodlighting in Europe and the United States. The lighting of the Statue of Liberty demonstrated both the problems and the possibilities. An 1886 news report read: "The attempt to throw beams of light upward, while effective so far as the pedestal is concerned, did not bring the statue into view on account of its darker color. It is very doubtful whether this can be done by electric lamps."[8] Incandescent tungsten filament lamps of some 20 million candlepower were in use at the statue by 1920, and the torch was refit with yellow glass backlit by a lighthouse lens that projected an intense beam outward.[9] Critics remained, pointing out the strong shadows cast under the statue's chin and around its eyes. In 1976, the lighting was enhanced, the statue itself lit by metal halide lamps providing four times the previous luminance. High-pressure sodium lamps rich in yellow were placed inside the torch. Mercury vapor lamps were positioned to cast a greenish-blue light on the statue's crown, while the pedestal was bathed in a blend of yellow sodium and white halide light to bring out the color of the granite.[10]

FIGURE 9.1. Lincoln Memorial, Washington, D.C., 1998. Sophisticated floodlighting highlights the building's various structural elements, giving them a distinctive nighttime relationship. The Lincoln Memorial exemplifies the contemporary perfection of lighting techniques first developed a century ago.

During World War I the Capitol dome in Washington, D.C., was lit, using floodlamps placed in clusters at the corners of the House and Senate wings. Wrote one observer: "Against the somber shadows of night, at this critical moment in our history, the inspiring white dome of our Capitol at Washington, high above the Federal City, stands resplendent in rays of shining light—a radiant monument to freedom and democracy."[11] So also were the various memorials in the city floodlit, to amplify their special architectural features at night. The Lincoln memorial is lit today to silhouette its massive columns (fig. 9.1).

The lighting of Niagara Falls provided a special challenge for lighting engineers. In 1879, a Brush arc light system of eighteen lamps was installed near the American Falls, only to be removed when the State of New York took ownership of Goat Island and its surroundings.[12] In 1884, in order to photograph the falls, artist Albert Bierstadt lit the same area by exploding charges of gunpowder.[13] In 1897, acetylene locomotive headlights were installed, casting enough light to make the American falls a nighttime attraction.[14] In 1908, Walter D'Arcy Ryan, with the backing of his employer General Electric, undertook the task of floodlighting the entire falls. A journalist for the *New York Tribune* reported: "Magnificently illuminated, the Falls were of a beauty that their daylight aspect has never equaled. For the first time since a factory was erected to draw its power from the rushing waters the garish outlines of the bleak brick buildings were gone, and in their place . . . were the falls in their old glory." He continued, "The sordid sight of the factories and the hurly-gurdies of the hotels and restaurants were banished. Presently the whole great stretch of the Falls was a mass of color; the whirling water beneath was like a pool of flame in the glow of the red searchlights."[15]

Not only were the falls "a riot of glorious beauty, so new, so strange, so marvelous—so like some unearthly and unexplained magic," as one visitor put it, but their nighttime floodlighting also served the important function of visual editing.[16] As historian David Nye emphasized, selective floodlighting enabled the landscape to be "edited, simplified, and dramatized."[17] The beautiful could be singled out and enhanced, while the ugly remained obscured in the darkness. At Niagara, the powerhouses, the hotels, and the other tourist facilities were obscured at night, and one could imagine the falls in some earlier, pristine state. From the 1920s on, Niagara Falls was regularly lit as a nighttime attraction in the summer months. A visitor from England found the effect "one of amazing and dazzling beauty." The lights ran "through a vividly changing scheme of colour: white, blue, orange, red, mauve, green," he wrote. "The white turns the rushing waters into a glistening marble veined with its own green . . . the other colours produce shimmering rainbow effects of amazing intricacy and beauty."[18]

SKYSCRAPERS

Commercial office buildings quickly proved to be the main consumers of floodlighting. Floodlighting could make an entire building shine in the night like a sign, helping define a prestigious address and make it attractive to tenants. Floodlights could be mounted on the roofs of adjacent buildings or on light poles or pylons surrounding structures at ground level, or they could be concealed behind parapets at setback levels or on projecting brackets on building walls. Visual "editing" could also be useful with building floodlighting. Only the upper towers of skyscrapers might be illuminated. Porticos might be lit or columns silhouetted, or other architectural features might be emphasized against the darkness. In 1908, arc light projectors were used to illuminate the top of New York City's Singer Building. Then the upper thirty stories of the Woolworth Building were lit, the tower's cream terra-cotta made to glisten under the soft glow of tungsten.[19]

Chicago's Wrigley Building was sheathed in white terra cotta deliberately chosen for its light reflecting capabilities, and the sheathing became progressively whiter (and hence more light-reflective at night) with increasing height. The building was described by one English visitor as "a veritable jewel at night that stands out against the dark sky like a sentinel clothed in snow-white, guarding all approaches to the city."[20] Eighty-six projectors located on adjacent buildings and forty-three projectors on the building itself generated some 20 million candlepower. The building could be seen at night fifteen miles out into Lake Michigan.[21] The nearby tower of the *Chicago Tribune* likewise was brightly illuminated from top to base, like a gigantic incandescent version of the Westminster Victoria Tower. Below, the streets were brilliantly lit, "showing toy motor-cars and . . . men and women walking like ants at the base of enormous cliffs."[22] Lighting engineers soon learned that when entire buildings were lit, the illumination at the top should be at least twice that of the bottom. Otherwise, when viewed from the street, the illumination appeared uneven.

Floodlighting in cities, like street lighting, was drastically curtailed during the Depression and World War II, languishing until the building boom of the 1960s and 1970s excited renewed interest. Then, many of the new international-style office buildings that came to dominate city skylines were boldly outlined. The new high-pressure mercury and high pressure sodium lamps were used, as were halide lamps. Modernism, expressed in the geometric simplicity of boxlike buildings, was well served. Beginning in the 1980s, postmodernism arrived to parody modernism's simplicity through an imaginative embrace of historical design allusions and traditional impulses in floodlighting (fig. 9.3). Additionally, glass-sheathed

FIGURE 9.2. Stouffer Hotel, Dallas, 1990. Sheathed in white stone, the building is boldly silhouetted against a night sky at the edge of downtown.

FIGURE 9.3. View from the John Hancock Building, Chicago, 1996. The tower of the AT&T Corporate Center is lit at night, highlighting selected structural elements in ways reminiscent of early-twentieth-century landmark illumination.

skyscrapers glowed in the dark, with their interior fluorescent lighting fully exposed to external view. Thus fluorescent lamps, little used in street lighting, nonetheless indirectly influenced the look of city landscapes at night.

SEARCHLIGHTS AND BEACONS

Americans used light to observe a new age of travel. Nothing enhanced a tall building's stature more than an aerial beacon, which demonstrated at once great height and a direct connection with modern America's fascination with flight. Perhaps the most celebrated searchlight was the Lindbergh Beacon, erected atop Chicago's Palmolive Building on North Michigan Avenue. A tribute to Charles Lindbergh and his solo flight across the Atlantic, the beacon was the gift of Elmer A. Sperry, developer of both a high-intensity electric arc lamp and the gyroscopic compass used in air navigation. The beacon cast two beams of an extraordinary 2 billion and 1.1 billion candlepower, respectively, through thirty-six-inch diameter parabolic reflectors. The lesser beam was kept pointed at the city's municipal airport, while the stronger beam revolved at two revolutions per minute and was visible for upwards of 100 miles. Some forty years earlier, Sperry had placed twenty electric arc lamps atop the old Chicago Board of Trade Building in a beacon of about 40,000 candlepower.[23] (Today, the world's most powerful searchlight, visible from space, beams a shaft of light straight up from the pinnacle of the pyramid-shaped Luxor Hotel in Las Vegas.)

In some cities, beacons or some other light display atop a prominent skyscraper were built to foretell the weather. In 1895, signal lights were placed on the roof of New York City's Manhattan Life Insurance Building on Broadway, 300 feet above the street. Electric arc lamps in various coded combinations indicated expected weather changes. Boston's John Hancock Tower had one of the most elaborate and most colorful weather beacons. Panels were backlit—blue for clear weather, flashing blue for cloudy, red for rain, and flashing red for snow.

Lighthouses continued to operate near coastal cities to aid in ship navigation. By 1907, the federal government had placed in operation 1,495 lighthouses and automated beacon lights and sixty lightships.[24] Electric arc lamps proved unworkable in lighthouse lenses, as the carbons tended to burn down and throw light beams out of trajectory. Thus oil lamps continued to serve lighthouses long after that technology had been replaced in other types of illumination. The lighthouse, however romanticized in novels and pictorial art, played only a small role in how night was experienced directly by the vast majority of urban Americans. More commonly known were the small battery-operated harbor lights of red and green marking navigation channels in ocean and lake ports.

Some cities used a large number of mobile searchlights, the lights becoming an important symbol of place. Los Angeles is the prime example. In 1928, searchlights of over 4 billion candlepower were assembled on Hollywood Boulevard for a light festival, sending a spectacular light display into the sky. Searchlights soon became a necessary adjunct to Hollywood premiers, helping to showcase new movies and their stars. The Otto K. Olesen Electric Company became Hollywood's prime purveyor of searchlights, capable by 1931 of putting seventy portable lights into action on any given night.[25] The searchlights carried strong theatrical connotations—the brightness of the "limelight" brought out-of-doors made entire landscapes into stages.

AIRPORT LIGHTING

Railroad and factory yards were floodlit, often with extensive sheets of light spread over large areas. Downtown in big cities, construction sites glowed in the night as excavations were dug and steel skyscraper skeletons erected by night as by day. But no commercial use of projected light was quite as impressive as airport lighting. By 1932, there were 630 municipally owned airports in the United States, 30 percent of them lit for nighttime operation. Of the 129 commercial fields, 19 percent were lit.[26] The night landings and takeoffs of scheduled mail planes in the 1920s became an attraction in many places, some cities holding light shows (or "Airport Nights") in conjunction with nighttime aviation. By 1940, larger airports were equipped with a variety of lighting fixtures: beacons, boundary lights, range lights, contact lights, obstruction lights, and lit wind cones, as well as floodlit hangar aprons and loading areas.

In 1931, Birmingham's Municipal Airport was equipped with a 2.7 million-candlepower rotating beacon located on a seventy-foot galvanized

FIGURE 9.4. A hypothetical view of nighttime airport lighting. Early use of beacons, boundary lights, hangar and apron lights, searchlights, and floodlight projectors for field illumination are depicted here in a painting commissioned by General Electric. From "The Lighting of Airports," *American City* 38 (March 1928): 104.

steel tower. Immediately above was a green auxiliary beacon flashing out "BM" in international Morse code. The standard at such airports was a thirty-six-inch, double-end beacon equipped with either a 500- or 1,000-watt incandescent filament lamp, showing six clear flashes per minute. Flashes of green from the auxiliary beacon filled the intervals.[27] These devices were intended to guide pilots to the airport and help them identify the facility. On the transcontinental air routes, beacon lights at twelve- to twenty-mile intervals flashed in alternating white and red. The 320-acre airfield at Birmingham was clearly outlined with white boundary lights placed low to the ground at 300-foot intervals.

Most large airports, as at Birmingham, had paved runways by the end of the 1920s, with banks of green range lights marking their end points. At some airports, yellow and white contact lights outlined the runways at 200-foot intervals, their luminaires turned toward incoming aircraft. From a distance, pilots could see glowing white dots outlining landing strips, with yellow dots toward the ends of runways. Sodium vapor lamps were widely employed for the yellow markers, incandescent filament lamps for the white. Obstruction lights were red, and they were placed on any structure within an airport's glide path. A standard twenty to one ratio was established for determining which structures would require obstruction lighting, a thirty-foot tower, for example, requiring lights if it was within 600 feet of an airport runway.[28] Because of the proliferation of both airports and tall buildings, blinking red lights (and flashing strobe lights in later decades) came to characterize night horizons almost everywhere in the United States.

Some airports, such as Detroit's City Airport, floodlit runways as well as their hangar apron and terminal loading areas. At Detroit, banks of floodlights with diffusing lenses to spread light beams down and out were placed at the ends of the runways.[29] By 1940, large airports were adding blue taxi lights to outline taxiways and traffic lights and directional signals to manage aircraft movement on the ground, in addition to lighting wind cones. Before the widespread use of radio and radar, at night pilots were totally dependent on lights in appraising airport landing conditions. A standardized lighting vocabulary was adopted nationwide. Often the names of cities were written in bold letters on hangar roofs and floodlit at night. Many airports maintained a "ceiling projector," a narrow beam of light pointed upward to give pilots a means of gauging cloud height. With all these additions, nighttime aviation brought a diversity of light and color to the American urban night.

After World War II, airport lighting became more and more sophisticated. Chicago's O'Hare Field in 1957 averaged 390 takeoffs and landings per hour. There were six tangential runways, each 8,000 feet long and 200 feet wide, with connecting taxiways leading to scores of passenger loading gates.[30] High-pressure sodium vapor lamps were widely used because of their high visibility and low cost of operation; left unscreened, they were ideal for outlining runways. Mercury vapor lamps with blue globes were used on taxiways, and fluorescent lighting was employed extensively inside the O'Hare terminal. From the 1950s on, airport terminals glowed at night, allowing—and encouraging—nighttime travel.

STADIUM LIGHTING

Floodlighting, used earlier to light bathing beaches, was applied to lighting playgrounds and athletic fields as well. By 1910, stadiums in Cincinnati and Chicago had been lit for lacrosse, soccer, and amateur baseball, using portable lights placed at ground level beside playing fields. Professional baseball was first played at night in Des Moines in 1930, when the local Demons of the Western League defeated the rival Wichita Aviators thirteen to six in a stadium floodlit with six ninety-foot towers, each tower holding eleven 1,000-watt incandescent filament lamps. "The whole affair was a 'lawn party,'" quipped the *Des Moines Register,* as "the only thing lacking for atmosphere was the Japanese lanterns." The lights were "a glaring success."[31] Reporters also noted the light-gray haze of cigarette and cigar smoke that floated up and out of the stands, shimmering in the light. The game, hailed as a historic nighttime innovation, was broadcast on the radio nationwide on one of the NBC networks.

Minor league baseball was first to take advantage of night lighting. "From a small beginning in the Prairie States beyond the Mississippi, night baseball has spread like a veritable conflagration," confided the *Literary Digest* in 1930.[32] The Pacific Coast League in the West and the International League and American Association in the East led the way. The Buf-

FIGURE 9.5. Night baseball at Omaha's Rosenblatt Stadium, 1993. Floodlighting fostered nighttime professional baseball, first in the minor leagues and then in the majors.

falo Bison drew ten times their average number of afternoon spectators to night contests.[33] As general manager of the Reds, L. S. MacPhail brought floodlights to Cincinnati's Crosley Field in 1935, using light to bolster attendance for a flagging franchise. When MacPhail moved three years later to the Brooklyn Dodgers, Ebbets Field was also lit. Night lighting then spread to Cleveland, Chicago's Comiskey Park, Philadelphia, and Pittsburgh. Amateur baseball was also stimulated by nighttime lighting, as many cities and towns illuminated municipal fields. Kewanee, Illinois, established a "twilight league" in 1931, with teams sponsored by the American Legion, the Elks, and the Odd Fellows. Attendance at games approached 5,000 in a town of 17,000 people.[34] After 1950, Friday night high school football games in brightly lit stadiums became a youthful rite of passage all across the United States.

FLOODLIGHTS, SPOTLIGHTS, AND SEARCHLIGHTS were used to place emphasis on nighttime landscapes. Their primary purpose was to attract and hold attention, making public landmarks stand out in the darkness. Commercial edifices were especially illuminated, not only owing to businesses' self-serving interests but also as a civic gesture, enhancing architecture during the night. Such pretenses were epitomized by tall office buildings topped by beacons. Just as floodlighting could be used to disclose, it could also be used to hide. By floodlighting one object but not another, landscapes could be visually edited at night. Government and big business were the important players in floodlighting practice, deciding what to emphasize in the night, just as the gentry had advanced their own social agendas with festivals and fairs. But, as at the amusement parks, the use of searchlights in places like Hollywood also heralded an evolving popular culture.

Applying light to athletic fields helped Americans fill their increased leisure time. The lighting of sports stadiums helped mature professional athletics—not just baseball but football, tennis, and horse and auto racing as well. Wherever outdoor play or work was extended into the dark hours, floodlighting was there, the lights themselves remaining an attractive force. Lighting up the night brought a sense of gaiety, enlivening activities and increasing attendance.

Street lighting substantially homogenized city streets, facilitating motoring, but lighting also held promise for other forms of transportation. In the twentieth century, the airplane was perhaps the one means of transportation that inspired more public admiration than the automobile. Lighting accommodated air travel in the night, as searchlights and beacons brought diversity to the darkened city—albeit also in standardized ways. The lighting of airports substantially fostered the maturation of America's aviation industry, eventually placing air travel on a twenty-four-hour schedule.

CHAPTER TEN

The Great White Way and Electric Sign Art

Perhaps nothing in the early twentieth century seemed more spectacular at night than the great "sky signs" perched high above city streets, blinking out commercial messages first in white light and then later in the colors of the rainbow. No street anywhere in America, or in the world, was more associated with sign spectaculars than New York City's Broadway, dubbed the "Great White Way." Electric signs did not originate in New York, however. London and Paris actually led the way, but on New York City's Broadway the use of electricity in outdoor advertising quickly rose to its most exaggerated form. Magnificent signs defined the street as an icon representative not only of the city but also of America, comparable to the city's skyline and its most important landmarks, such as the Statue of Liberty. The art of the electric sign, as perfected on Broadway, was replicated in other cities. On business streets in large cities and small towns, Broadway's bright lights inspired scaled-down equivalents.

The electric sign was adopted very quickly across the United States after 1910, and it was aggressively marketed by the electric utility companies. A small handful of firms came to dominate the making of electric signs, either through their connections with the utility companies or through control of patents for electrical circuitry, metal plating, or new glass and plastic materials. Electric circuitry controlled lamp operation; plated metal (and later plastic sheeting) withstood weathering and lengthened sign life; and glass and plastic lenses and sign facings diffused, intensified, or focused light. Large regional and even national companies evolved to market the materials for signs and, indeed, to market highly standardized finished signs. Most power companies maintained design departments to recruit customers, turning actual sign manufacture over to sign fabricators. Most of the electric signs seen in the nation's business districts or out along evolving commercial strips, certainly through World War II, were actually owned by the power companies and leased out on monthly or yearly contracts. Also contributing to standardization were city ordinances that governed sign technology through construction codes.

TIMES SQUARE

Until Oscar Hammerstein opened the Olympia Theater at Broadway and Forty-fourth Street in 1895, the area immediate to Times Square, then Longacre Square, had been nicknamed the "Thieves Lair," a seedy locale deemed dangerous after dark. But new theater construction long had been headed up Broadway, pushed by inflating property values to the south. Rudolph Aronson had opened the Casino Theater at Broadway and Thirty-ninth Street in 1882, which was followed by the Metropolitan Opera House at the same intersection a year later. The Broadway Theater at Forty-first Street opened in 1893.[1] Vaudeville rapidly followed the legitimate theater into the area; Hammerstein's Victoria Theater, for example, opened in 1904. The introduction of brightly lit theater marquees illuminated Broadway, making the street safer after dark. Until then, only a lone gas lamp had lit the square. The theaters brought bright marquees and illuminated signs, setting Broadway and Times Square literally awash with light.

Broadway's golden era lasted from 1900 until the stock market crash of 1929. The number of productions in theaters on or immediate to Broadway numbered seventy in the 1900–1901 season.[2] In 1917–18 there were 126 productions, and the number reached a high of 268 in 1928–29.[3] Not only did the Depression evaporate investment capital for the theater, but the motion picture industry continued to prove an increasingly successful competitor for the public's entertainment dollar, especially after talking pictures arrived. In the 1920s, many Broadway theaters were converted to movie houses or demolished to make way for lavish movie "palaces." New films were premiered not only in Hollywood but also in New York City at such theaters as the Paramount, the Rialto, the Roxy, and the Strand. With the rise of broadcasting in the 1930s, other theaters were converted into radio studios and, starting in the late 1940s, into television studios. Some remaining vaudeville houses struggled on as burlesque houses or rehearsal halls, while others closed. The last full vaudeville bill at the famed Palace Theater was in 1932.[4]

Hotels and restaurants reinforced Broadway's image as America's entertainment capital; the hotels helped serve the new Hudson River docks to the west, where the largest oceanliners berthed. The new Astor Hotel at Forty-first Street was advertised as the "most electrified hotel in the world" and was home to entertainers and artists such as George M. Cohan and Enrico Caruso.[5] Other notable accommodations included the Knickerbocker, the Normandie, and the Vendome. Fancy eateries like Churchill's, Murray's Roman Gardens, and Rector's helped highlight the street. Rector's was "the supreme shrine of the cult of pleasure" according to historian Lloyd Morris. "It was the American cathedral of froth and frivolity, the terminus of the prim rose path."[6] Opened in 1899 on the square, its dining rooms were decorated in the style of Louis XIV, adorned in crystal, floor-to-ceiling mirrors, and gold gilt to reflect opulently in incandescent light.

Broadway's roof gardens provided another important entertainment diversion, mixing the theater and the restaurant in novel setting. Elevated ten or more stories above the street, the roof gardens were vividly lit by electricity and provided bird's-eye views of a nighttime landscape, itself variously illuminated by oil, gas, and electric lights. The first roof garden opened in 1882 atop the Casino Theater, financed by several of New York City's wealthiest families as a summer concert hall. The Moorish architectural motif, highlighted by electric incandescent lamps, was a deliberate allusion to the Newport Casino in Newport, Rhode Island, then an elite resort of the highest order. The roof garden at Madison Square Garden, with electric searchlights atop its tower, brilliantly outlined in incandescent bulbs, opened in 1892. Variety acts were drawn from the music halls downtown and from the cafe–concert halls of European cities.[7] With increased affluence and leisure time in society generally, patronage of the many roof gardens increasingly became middle class.

At Oscar Hammersteins's roof garden atop his Olympia Theater, as at other rooftop gardens, the novelty of an elevator ride contributed to the appeal. Exiting the car at the top of the Olympia, patrons entered an immense glass enclosure across which some 3,000 electric incandescent lamps were festooned. Over the glass ran a constant stream of cold water, pumped up from refrigerator tanks in the building's basement. Plants hung from the iron columns, adding greenery. On a small stage performers entertained to a sea of straw hats obscured in clouds of cigar and cigarette smoke.[8] Writer Rupert Hughes rejoiced, "Best of all, they have roof-gardens in the summer, and there can be nothing more beautiful than to dine far above the world in the open air, under the stars, with the moonlit river in the distance and a spring breeze tugging at the napery and winnowing your soul."[9] The magnificent views impressed diners. "To the South . . . the city ran down to flaming towers in the glistening haze that seemed a luminous vapor rising from dazzling avenues," wrote Owen Johnson, setting the mood for a scene in his 1915 novel *Making Money*. "Looking down on myriad points of light one seemed to have suddenly come upon the nesting of stars, where planets and constellations took flight toward the swarming firmament."[10]

The lower portion of Broadway closed up and emptied out after sunset in the late-1880s, but above Twenty-third Street a vigorous nightlife thrived. "From Union Square to Thirty-Fourth Street the great thoroughfare is ablaze with the electric lights, which illuminate it with the radiance of day," wrote James McCabe. "Crowds throng the sidewalks; the lights of the omnibuses and carriages dart to and fro along the roadway like myriads of fire-flies, the great hotels, the theaters, and restaurants, send out their blaze of gas-lamps, and are alive with visitors."[11] In 1900, William Archer observed: "Theaters, restaurants, stores, are outlined in incandescent lamps; the huge electric trolleys come sailing along in an endless stream, profusely jeweled with electricity; and down the thickly gemmed vista of every cross street one can see the elevated trains, like luminous winged serpents, skimming through the air."[12]

Rupert Hughes found Broadway "one long canon of light. Even the shops that were closed displayed brilliantly illuminated windows. In some of them all the trickeries of electricity were employed and rhapsodies of color glittered in every device or revolved in kaleidoscopes of fire."[13] In 1913, Alan Raleigh found in Broadway "a pilgrimage through Dante's

Inferno." "Hurrying crowds shoulder one another off the pavement . . . there is an incessant clanging of bells and motor horns . . . the line of tram-cars is almost continuous . . . the sky, twinkling with stars, is seen far above the lofty buildings."[14] But to him the most "bewildering" sights were the giant "sky signs" that broke the darkness high above the street. There, a "Niagara" of light glittered most spectacularly.

BROADWAY'S SIGN SPECTACULARS

The first giant electric sign—the first sign spectacular—was erected in May 1892 high over Broadway on the north side of the Cumberland Hotel, adjacent to the site that would become the Flatiron Building a few years later. Covering some five stories, the sign was clearly visible to people coming south on Broadway through Madison Square. It advertised the upscale Manhattan and Oriental Beach resorts at Coney Island with the words:

SWEPT BY OCEAN BREEZES

MANHATTAN HOTEL

ORIENTAL HOTEL

SOUSA'S BAND

PAIN'S FIREWORKS

HAGENBECK[15]

Theodore Dreiser, remembering the sign as worded differently, recalled that "each line was done in a different color of lights, light green for ocean breezes, white for Manhattan Beach . . . red for Pain's fireworks and . . . and blue and yellow for the orchestra and band."[16] Each line was illuminated separately and then lit together before the sign repeated its cycle.

Within the year, the sign was leased to the H. J. Heinz Company and a forty-five-foot-long pickle outlined in green incandescent bulbs appeared even higher up on the hotel's mansard roof.[17] Critical voices were raised, especially after the Dewey Arch, commemorating American victories in the Spanish American War, had been erected in the square immediately below. The Heinz sign's electric glare at night illuminated the arch and its surrounding light pillars in ghostly shades of green, red, and orange, much to the chagrin of the Municipal Art League. "Advertising Run Mad," was the headline for an article in *Municipal Affairs:* the sign made Madison Square "unimaginable except in nightmare," the essay's author complained.[18] Now gentry taste rallied against the incipient rise of a commercially driven popular culture. As large signs became more numerous and spread northward along Broadway from Twenty-third Street, proposals were made to control the signs and to decorate the street with monumental "electroliers" instead, to be highlighted by an electric fountain in Times Square.[19]

An accumulation of signs along Broadway quickly evolved. William Archer, an English visitor to the city in 1900, could remark that the "flashing out-and-in electric advertisement which makes the night hideous in London" was "unknown in New York." "One or two steady-burning advertisements irradiate Madison Square of an evening; but being steady they are comparatively inoffensive," he wrote.[20] Yet only four years later Rupert Hughes reported: "From most of the buildings hung great living letters. Some of these winked out and flashed up again at regular intervals. Others of them spelled bulletins in sentences that flared automatically."[21] John C. Van Dyke wrote in 1909 that the electric signs "shown everywhere" and "one wearies unto death with what they say." They were located over doorways and on building facades, but the most conspicuous were perched on roof-tops. "Letterings, patternings, arabesques, figures of birds and beasts and men, are outlined by small electric globes, and the whole thrust upon the night in giant proportions." "All told," Van Dyke continued, "the glitter and glare of these signs make up a bewildering and (it may be admitted) a brilliant sight. Great throngs of people delight in them."[22]

Attempts to control advertising signs on Broadway had quite the opposite effect. The Broadway Association promoted signs vigorously, and in 1916 the new city zoning law permitted giant signs with relatively few restrictions. On fireproof buildings, signs could be erected up to seventy-five feet above the roof level, whereas on other buildings a height of only fifty feet was allowed. No sign could project more than eight feet from the side of a building.[23] This was not the case on Fifth Avenue, however, where the merchant's association had agreed to ban large signs of all kinds. Thus the two thoroughfares developed very different personalities, the vulgar advertising excesses of the one contrasting with the sedate "tastefulness" of the other. The two streets did complement one another, however, allowing people and atmospheres of very different status and class to exist in very close proximity. Fifth Avenue became the street of exclusive clothiers, jewelers, bookstores, and cafes and restaurants. Broadway, driven by the theaters, roof gardens, inexpensive restaurants (Rector's had closed in 1915), and wide array of tourist-oriented shops and arcades, brought a touch of the world's fair midway and Coney Island to Manhattan.

Broadway became New York City's leading tourist attraction. O. J. Gude, whose company pioneered the street's electric signs, coined the term *Great White Way*. In 1913, he wrote enthusiastically: "Sightseeing coaches nightly take strangers up Broadway to see this phantasmagoria of lights and electric signs. It is a great free exhibition for strangers from all over the country and Europe." Broadway was "one of the sights of America that paints its picture . . . vividly on the mind."[24] It was, according to Gude, one of the nation's three great "awe-inspiring" sights— along with "the gigantic skyscrapers" of Manhattan and Niagara Falls. Europeans tended to agree. Pierre Loti, visiting from France, wrote: "Everything seems to vibrate, to crackle, under the influence of these innumerable currents which dispense power and light. One is himself electrified almost to the point of quivering under the stimulus." In comparison, Paris, he wrote, "will seem just a quiet, old-fashioned little town."[25] As for New York, "It was very strange and even a little diabolical; but it is so droll, so ingenious withal, that I am immensely amused and even on the verge of admiration," he admitted.[26]

More extraordinary signs were erected. Atop the Normandie Hotel at Thirty-eighth Street, Elwood Rice created an illuminated Roman chariot race seventy-two feet high and ninety feet wide. Some 20,000 incandescent bulbs created the illusion of galloping horses, straining drivers, spinning wheels, and snapping whips.[27] During the first week it was displayed, enough crowds gathered to halt traffic. Above the race, the names of corporate sponsors were illuminated in sequence. A toothbrush ad appealed to Rupert Brooke:

Out of the gulfs of night, spring two vast fiery tooth-brushes, erect, lean-
ing towards each other, and hanging on to the bristles of them a little Devil,
little but gigantic, who kicks and wriggles and glares. After a few moments
the Devil, baffled by the firmness of the bristles, stops, hangs still, rolls his
eyes, moon-large, and, in a fury of disappointment, goes out, leaving only
the night, blacker and a little bewildered, and the unconscious throngs of
ant-like human beings.[28]

Here on these signs were the moving cartoon characters in anticipation of
Hollywood animation.

The signs of Broadway were most concentrated at Times Square. There,
a virtual "who's who" of aspiring corporate names lit the night: Wrigley's
chewing gum, Budweiser beer, Warner's Rust Proof Corsets, Corticelli silk.
The Cliquot Club Ginger Ale sign at Forty-first Street was fifty-five feet
high and sixty-five feet wide, and it contained 3,000 incandescent bulbs
colored green, yellow, purple, red, orange, and white.[29] Firms like the Federal Sign Company bought electricity wholesale from New York Edison,
leased space on building facades and roofs, and sold their design and engineering services to such customers as Cliquot Club. Many buildings in
Times Square were completely covered by advertisements, their upper stories rendered useless as office space. Indeed, the 225-foot Heidelberg
Tower was built solely to serve as a platform for signs.[30]

The advertising art of Broadway gained considerable respect as cultural
icon. *Atlantic Monthly* called electric signs "our folk-art writ in fire in the
sky."[31] Signs stimulated art. Vachel Lindsay wrote "A Rhyme about an
Electric Advertising Sign," which began:

> *I look on the specious electric light*
> *Blatant, mechanical, crawling and white,*
> *Wickedly red or malignantly green*
> *Like the beads of a young Senegambian queen.*
> *Showing, while millions of souls hurry on,*
> *The virtue of collars, from sunset till dawn.*[32]

Visitors to New York City sought out Times Square as the symbolic heart
and soul of the city. In the evening, wrote Louis Dodge, the "Doctor Ken-

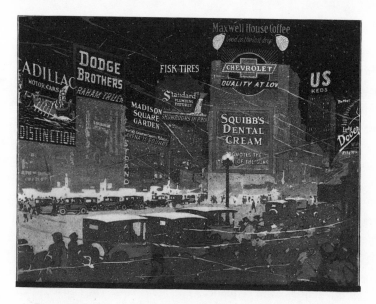

Broadway—*a National Advertising Medium*

Circulation a million people a day.

Two million prosperous visitors a month from all the states of the Union.

Reader interest so high that spectacular electric advertising on Broadway is talked about thousands of miles away.

Cost per thousand circulation averages 28 cents.

Details as to locations, costs and designs will be promptly furnished.

General Outdoor Advertising Co.

One Park Avenue
New York   *Sales Offices and Branches in 60 other Cities*   Harrison & Loomis Sts.
Chicago

FIGURE 10.1. A General Outdoor Advertising Company advertisement. From *Poster* 19 (Feb. 1928): 27.

nicotts of the inland towns walk their more or less restless womenfolk, and try to believe they have struck oil in an emotional sense, and that they are having the time of their lives."[33]

Broadway generated the highest traffic volumes in the city. By the late 1920s, an estimated one million people made their way through Times Square daily (fig. 10.1). In the immediate vicinity there were nearly one hundred theaters capable of seating 9 million people a month. There were ninety-four hotels within walking distance, which could accommodate 2.5 million guests a month.[34] Signs changed season to season and year to year, as few advertisers could afford the rapidly escalating rents indefinitely. By the time the Wrigley Company took down its "Spearmint" sign, the seven years it had been displayed had cost the company over $700,000.[35] Rents for sign space were as high as $10,000 a month for prime locations, and the costs of maintenance were also exorbitant. Before the era of automated circuits, each large sign required a modest crew of laborers who manually operated the switches, and, even after signs were automated, workers were needed each night for systematic bulb inspection and replacement.

New kinds of signs appeared. The "talking sign" wrote out words across a matrix of bulbs, some of which were lit in various combinations to spell out sentences in sequence. The New York Times used such a device to announce the results of the 1928 presidential election, when Herbert Hoover defeated New York's Al Smith. The 14,800 amber bulbs in the board could produce 261,925,664 flashes an hour.[36] The New York Times sign was used until 1978, long after the newspaper had vacated its Times Square building.

The 1920s saw the rise of giant floodlit "billboards," as easily read in the daytime as at night. At that time, billboards were painted rather than laid on as paper "hoardings." The entire facades on several theater buildings were covered by billboards, beginning in the 1920s, advertising one or another motion picture. Cecil B. DeMille's The Ten Commandments commanded attention when it appeared on a forty-foot-high sign on the Criterion Theater. The chariots of the Pharaoh trapped by the Red Sea and a giant Moses holding his tablet were depicted. Warner Brothers' The Gold Diggers of Broadway prompted a 210-foot sign across its Hollywood Theater, stretching, in essence, for an entire city block from Fifty-first Street to Fifty-second Street.[37] The big innovation of 1927 was the introduction of Claude tubes, or neon, a development that would flower in the 1930s.

In 1936, Wrigley returned to Broadway with a new sign deliberately designed to be the world's largest. The sign faced the Astor Hotel on the east side of Times Square between Forty-fourth and Forty-fifth Streets. Seventy-four feet tall and 188 feet long, the display weighed over 110 tons. Its central feature was a gigantic multicolored fish that appeared to glide rhythmically in waves of sea-green light. Bubbles rose from the bottom to the top of the sign. The messages "Steadies the Nerves," "Aids Digestion," and "The Flavor Lasts," were repeated in sequence. The Wrigley sign used 1,084 feet of neon tubing and 29,500 frosted incandescent bulbs.[38] Also in 1936, a new "Planters Peanut" spectacular, fifty-five feet tall and forty-nine feet wide, appeared at the north end of Times Square, along with giant Four Roses, Chevrolet, and Coca Cola signs arrayed on the same building. Approximately 6,700 incandescent bulbs and 3,340 feet of neon tubing were used in the Planters sign.[39] Red tubing spelled out "Mr. Peanut Greets You" in ten-foot letters. In 1941, the Wrigley sign came down. In its place a huge Bond sign was built, advertising the world's largest men's clothing store, installed below in a totally remodeled building. The structure's streamline modern design was amplified through use of neon.

Most Broadway signs were dismantled during World War II, but after the war vacated spaces once again filled with mammoth advertisements. In 1947, Charles Le Corbusier, the French architect, called Broadway "that incandescent path cutting diagonally across Manhattan." "Electricity reigns," he wrote, "but it is dynamic here, exploding, moving, sparkling, with lights turning white, blue, red, green, yellow." It was a "nocturnal festival characteristic of modern times."[40] Yet incandescence was not what dominated Broadway's illumination. Especially in Times Square, neon—and even the new fluorescent tubes—had come to dominate. Francis Marshall wrote in 1949, "Sheer vulgarity, if on a big enough scale, can become an art. You are bludgeoned into admiration."[41]

In the 1950s, the floodlit billboard was joined on Broadway by signs consisting of plastic screens lit from behind by fluorescent tubes. The idea was to increase levels of luminance above the capabilities of floodlighting while keeping construction and operating costs low. To achieve some sense of spectacle, these signs often used neon or incandescent lamps for highlighting slogans or outlining. Large billboards were still used to promote motion pictures, and many were elaborated by features that stood out, giving a sense of three-dimensionality. A giant Viking ship, for example, promoted the *The Vikings* with Kirk Douglas and Tony Curtis. Ornamented movie marquees with backlit panels contributed to the sense of luster, but the new billboards tended not to blaze so brilliantly in the night and few were animated.

There were relatively few truly spectacular signs in the 1950s and 1960s. Several used new prismatic materials that were lit from behind or floodlit or both. Hiram Walker and Son's new Canadian Club sign, created by the Artkraft Strauss Sign Company, used red prismatic letters twelve to twenty-two feet tall. The hollow letters were three feet deep, their front edges extending about ten inches out from the yellow porcelain enameled metal background, itself divided into a checkerboard by yellow fluorescent tubes arranged in patterned sequences across the grid. The protruding letters were outlined by orange incandescent bulbs, and the hollow insides and flanged fronts of the letters were highlighted with ruby red neon.[42] The sign was designed to give different impressions when viewed from different angles and to be vivid in the day as well as in the night.

Movable screens were also popular beginning in the 1950s. Backlit panels on signs could rotate to display different written messages or pictures. The Trans World Airlines sign erected at Forty-third Street featured a two-fifths scale model of a Lockheed Constellation complete with turning propellers and running lights. The sign was seventy five feet high and 100 feet long, the twenty-foot letters of its "Fly TWA" message filled with incandescent bulbs and outlined in neon. Its "Scenerama" stage measured twenty by thirty feet and was mounted on four turntables so that scenes from cities on TWA's routes could be rotated at intervals. The sign used 20,000 incandescent filament lamps, a mile of neon tubing, and over fifty miles of wiring.[43]

Broadway's personae changed. Within the regular year-to-year turnover of signs, there was a drift toward less electrical brilliance and more floodlit or backlit billboards (fig. 10.2). In 1955, the giant Bond spectacular came down in favor of an equally large Pepsi Cola sign featuring two glowing fifty-foot Pepsi bottles and the slogan "The Light Refreshment," the whole silhouetted against a floodlit background. It took its turn as the largest of the electrical signs on Broadway and included a waterfall system lit from behind. Immediately below, wrapping around the corners of the building, ran the CBS "Telesign," across which raced news reports. It employed modern electronic circuitry and was a substantive improvement

FIGURE 10.2. New York City's Times Square, 1976. The brilliant neon and incandescent lights are mostly gone, replaced by floodlit billboards and backlit panel signs, especially on the remaining theater marquees. Electric light entertained primarily at or near street level.

over earlier "talking signs" or "traveling message panels." At street level, the Bond Clothing Store was all but obscured by a maze of neon and back-lit acrylic-paneled signs. The radiance at street level all around Times Square, heightened by new, intense street lighting and backlit movie mar-quees outlined in neon, made up Broadway's new light show. By 1970, the glittering, flashing, glaring brilliance of Broadway was more evident on the street and sidewalk than on rooftops.

New high-rise office buildings and hotels intruded on Broadway in the 1970s and 1980s, replacing many of the small theater buildings upon which signs traditionally had been placed. Critics decried the walls of glass and steel, devoid of signs, and in 1985 the Mariott Marquis Hotel agreed to have four signs placed on its front facade. These signs had to hang on the building's wall, so developers sought lighter materials and new ways of projecting light that would not involve heavy apparatus. Kodak's sign was a thirty-by-fifty-foot box covered by a huge fabric and backlit by fluo-rescent tubes. Exhaust fans prevented heat from building up inside and also kept the fabric taut. Beneath the picture displayed on the screen was an electronic "crawl sign" (a new name for the telesign), across which the company's products were promoted with sequenced slogans. A special Midtown Zoning District was established in 1982 in an attempt to preserve the "Great White Way" as a district for outdoor advertising spectaculars. Stipulations required developers of five new skyscrapers, planned between Forty-second and Fiftieth Streets, to include a minimum of 60,000 square feet of total sign space.[44] A renewed infusion of neon was encouraged on what had become very much a street of illuminated billboards and glaring street lights.

BROADWAY'S DECLINE AND REBIRTH

The image of the "Great White Way" was indelibly fixed on the public's eye in the 1920s. Yet even then the seeds of decay had already been sown. Little by little, Broadway and Times Square declined both as an entertainment district and as a zone of spectacular signs. Laurence Senelick called the process the "sleazification" of Times Square.[45] The Volstead Act of 1919, banning the sale of alcoholic beverages, started the process. Deprived of liquor sales, the restaurants and roof gardens closed. With the coming of prohibition, "the old revelry of the Great White Way shut down," commented the *Literary Digest* as early as 1922. Broadway was left "in dull, sparkless ruins." The street had become "a mere Main Street of motion-picture emporiums and of synthetic orange juice booths."[46] From the freely accessible roof gardens, New York City's nightlife descended into the closed basements of the speakeasies. Cheap shops, the kind of establishments that crowded Surf Avenue in Coney Island or Atlantic City's Boardwalk, squeezed into spaces where respectable restaurants and cafes had once thrived.

The number of legitimate theaters steadily declined over the decades. During the 1950–51 season there were eighty-one productions, falling to sixty-two in the 1969–70 season.[47] The veneer of seediness hardened with the arrival of peepshows and pornographic bookstores in the 1960s. The sidewalks of Times Square were crowded with prostitutes and panhandlers, attracted to the crowds which were, in turn, attracted to the lights. Although diminished, the lights still symbolized the very heart of the city, if not the city itself. More than ever, the lighting was a grand illusion, propped up and artificially sustained; here was an icon of city and a nation, too ingrained to let evaporate. Nonetheless, as an Australian visitor wrote: "Stand for a moment in Times Square . . . and watch a giant mechanical face blow smoke-rings in the air on behalf of Camel cigarettes. You couldn't feel lonelier anywhere."[48]

It was the anonymity of Times Square that attracted the underclasses. Pass the striptease joints, the sex cinemas, and the massage parlors, and you also passed con men, drug dealers, pickpockets, and sailors on the prowl. Men outnumbered women in Times Square three to one in the daytime and seven to one at night.[49] "You see, too, the gaudy hoardings which promise cinematic visions of torrid love and naked abandon. And despite the tumbling electrical waterfall and whirling, neon-lit aeroplane, despite all the enticements to buy this crazy gadget or that vicarious pleasure, the center of the world nearly always looks like a gathering of lost souls."[50] "Debris is strewn over the pavement and in the gutters," wrote McCandlish Phillips in 1974. "And wind stirs up scraps and creates a funnel of newspaper that swirls up in a mad ballet."[51]

The idea of Times Square as a zone of giant, glittering advertising signs found new life in the 1990s. Overturning plans for a thirteen-acre urban renewal project financed through the Empire State Development Corporation which would have converted the area into a sanitized zone of giant office towers, the city, its planners, preservationists, and developers reached consensus over a different kind of future, one in which a new generation of sign spectaculars would take center stage. Instead of wholesale demolition and rebuilding, many old structures—especially theaters—would be retained and converted to new entertainment uses. Several could be made into indoor theme parks, one of which would be run by Disney. Things had come full circle: Coney Island developers had followed the Great White Way in their use of night light. Walt Disney had mimicked Coney Island in developing the modern theme park. And now theme park aesthetics stood to help revitalize Times Square as "the Crossroads of the World."

EXTENDING THE GREAT WHITE WAY NATIONWIDE

Although sign makers operated within highly controlled economic and political environments, they did produce in most cities very lively nighttime scenes which, at a superficial glance, suggested a diversity like that of Times Square. This was especially so of principal shopping streets in downtown areas. Part of the apparent variety involved signs that were not fully "electric" (that is, built around electrical lamps emphasizing bright luminance) but merely illuminated by electric light. Examples included illuminated billboards and wall signs painted on downtown buildings. Electric signs, as most Americans thought of them, were intended to be a nighttime phenomenon, like the illuminations of New York City's Great White Way. Their purpose was to stand out vividly in the darkness and dazzle the eye through overt display of electric lamps.

Outdoor advertising differed from other advertising media. Signs did not circulate their messages to a market; rather, the market circulated around them.[52] Messages were aimed at audiences who were for the most part "other-directed." That is, signs were oriented toward people moving through a landscape to some destination. Impressions had to be communicated quickly, as passing motorists might be exposed to a sign for only a few seconds. Messages not only needed to register quickly, but they also had to be visible at a distance through the confinement of automobile windshields. At night, electric signs and electrically illuminated billboards were absolutely vital to outdoor advertising. While darkness obscured much of the surrounding landscape, electric light focused viewer attention—the more vivid the display, the more riveting the focus. It was difficult for people to ignore or "edit out" electric signs in the darkness, especially when signs were placed in the most visible locations. As critics of signs complained, they could not be turned off at will. Whereas one could turn aside a newspaper or magazine ad and tune out a radio commercial, electric signs were simply too intrusive on the nighttime landscape to be ignored.[53]

The role of advertising was to promote through repetition. When combined with advertising of other forms, the outdoor sign helped stimulate memory, creating a sense of familiarity with—and even trust for—a business or a brand name. Advertisers wished to engender a sense of con-

FIGURE 10.3. A cartoonist's view of the world of outdoor advertising in the 1920s. Signs that energized Broadway could be used in every city. Outdoor advertising made the world more desirable, the cartoon argues, to pedestrian and motorist alike. From Thomas Elwyn, "Signs—The Untiring Force Which Keeps the Wheels of Progress Turning," *Signs of the Times* 51 (Nov. 1925): 5.

stancy, a feeling that the product or service endured over time. Therefore, the building of a cumulative effect was essential.[54] The flashing electric light—the rhythmic turning on and off in the dark—did, itself, through repetition, hammer a message home. When used, pictorial effects were cartoonish, highly suggestive in outline but lacking in detail, and typically displayed the same repetitive action.

ELECTRIC SIGN ART

The vast majority of America's electric signs were of very modest scale. Through the 1920s, the following types predominated: (1) panel signs with exposed lamps, (2) glass globe or "canteen" signs, (3) floodlit reflecting signs, (4) enclosed panel signs with backlit lenses or faces, (5) open silhouette signs that were backlit, (6) floodlit billboards, and (7) projected signs. Incandescent filament bulbs or luminous tubes (neon or fluorescent) usually provided the luminance. Exposed signs with incandescent filament lamps numbered an estimated 265,000 in the United States in 1929, while there were about 120,000 enclosed signs with incandescent bulbs and 160,000 floodlit billboards.[55] Neon was just coming into use, and fluorescent technology was yet a decade away, but the electric sign industry was already mature. Firms such as Chicago's Federal Sign Company, a derivative of Commonwealth Edison's lamp leasing operation, were already operating nationwide, renting porcelain enameled panel signs with neon tubing. Trade associations like the National Association of Electric Sign Manufacturers and the Associated Sign Crafts of North America were also active.

In Chicago, the number of illuminated signs mushroomed from 850 in 1902 to 7,250 in 1912.[56] Although the city really never developed a show street of concentrated sign spectaculars like New York City's Broadway, large and small electric signs were distributed throughout the city, encouraged by Commonwealth Edison's vigorous salesmanship. The company furnished, installed, and maintained electric panel signs of standard blue-enamel for small retailers. New York Edison, through its Electric Sign Bureau, likewise promoted illuminated signs throughout New York City. Sales representatives approached commercial customers with sketches of prospective signs, then submitted the chosen options to sign fabricators. The company applied for city permits, obtained financing, and arranged for installation, prorating sign costs in monthly power bills to the customer.[57] In 1907, there were an estimated 10,000 electrically lit panel signs in all of New York City.[58] By 1924, there were an estimated 9,500 illuminated signs and one million incandescent lamps in Manhattan between the Battery and 135th Street alone. Of course, Broadway, with its 1,360 signs containing 350,000 lamps, was brighter than any of the city's other thoroughfares.[59]

William J. Hammer is credited with creating the first electric sign. In 1882 at the International Electrical Exhibition in London, he installed a panel with the word *Edison* spelled out in exposed standard 16-candle-power incandescent bulbs.[60] The first automated flashing sign was displayed a year later in Berlin. An 1884 sign in Boston spelling out "Boston Oyster House" is thought to have been the first electric sign for perma-

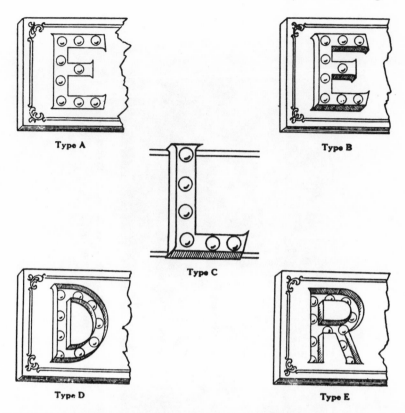

FIGURE 10.4. Design variations for the panel sign with exposed lamps. From Leonard G. Shepard, "Sign Lighting," in *Illuminating Engineering Practice* (New York: McGraw-Hill, 1917), 540.

nent use in the United States.[61] However, it was not until 1909, with the introduction of tungsten filament incandescent bulbs of low candlepower and low wattage, that the exposed-lamp electric sign, and, indeed, electric signs generally, won widespread acceptance. In 1911, development of the 2.5-watt, 10-volt Mazda "sign lamp" sped the adoption of electric signs nationwide. Signs consumed far less electricity than before and required less maintenance, thanks to a bulb life expectancy of some 2,000 hours. The new lamp also allowed signs to be wired in parallel rather than in series, thereby eliminating complicated systems of modified-series wiring.[62]

The panel sign with exposed incandescent lamps came in a variety of forms: (1) flat panels studded with lamps, (2) raised letters outlined with

FIGURE 10.5. A panel sign with rear-lit studded lettering. Such signs typified small storefronts through the 1920s. From George How, "Modern Business-Getting Methods," *Electrical Age* 37 (July 1906): 6.

lamps, (3) letters with lamps displayed on a skeletal framework, (4) letters sunken in troughs with lamps raised in the troughs, and (5) letters sunken in troughs with lamps additionally recessed (fig. 10.4). Flat panel signs with standard backlit studded lettering were easily assembled from prefabricated materials. Messages on panel signs tended to be short and to the point; words like *eat* and *cafe* identified restaurants, for example. Often a panel sign merely carried the name of an establishment's proprietor. The idea was to attract attention and drive home a clear message as invitation.

Los Angeles, perhaps owing to its distance from New York and other eastern centers, was slow to adopt signs lit by incandescent bulbs. In 1912, New York City had about 7,000 electric signs but Los Angeles had only thirty-three.[63] After World War I, however, with the city's economic boom, Los Angeles became a competing center of sign art innovation. In 1924, on a hillside above the movie capital, a developer erected a landmark sign studded with lamps spelling out "Hollywoodland." (The sign was allowed to deteriorate and then was renovated in shortened form—producing the "Hollywood" sign that stands today.) Many cities erected slogan signs, usually at prominent hillside, riverfront, or rooftop locations. Among the more memorable were: "Pittsburgh Promotes Progress," "Shreveport Spells Success," "Watch Wichita Win," "Dwell Here and Prosper" (at Allentown in Pennsylvania), and "Biggest Little City on Earth" (at Parkersburg in West Virginia).[64]

The small, flat panel sign lent itself well to neon tubing, which rapidly diminished the use of exposed incandescent bulbs. Whereas Los Angeles had been slow in adopting signs with incandescent lamps, the city was quick to embrace neon. Indeed, neon literally spread eastward across the United States from Southern California.[65] The wide adoption of neon in the West was encouraged by the region's use of alternating current gauged at sixty cycles per second. In parts of the eastern United States and eastern Canada, current was cycled as low as twenty-five cycles per second, which put an irritating flicker into neon tubes.[66] When key patents (especially one for a long-lived electrode used with neon tubing) expired in 1932, the manufacture of neon signs skyrocketed everywhere. As World War I had done, World War II brought about a dimming of electric advertising signs. After the war, fluorescent tubing came into prominence, displacing neon just as neon earlier had displaced exposed incandescent filament lamps in outdoor advertising art.

When neon first arrived it was trumpeted as "the light of the future."[67] Although neon was used in large sign spectaculars on Broadway and around New York Harbor (the Sunshine Biscuit sign, for example, was re-worked in neon tubing), it was more widespread among small business advertising. In 1926 Nathanson's Used Car Exchange at Broadway on Sixty-Second Street erected modest neon signs that spelled out "Nathanson" in orange and "Used Cars" in green.[68] Other early adopters in New York City included Perlman's Pianos on Grand Street, the Gold Seal Laundry on Atlantic Avenue, the National Cloak and Suit Company on West Fourteenth Street, and all of the city's many Klein Rapid Shoe Repair Shops. An advertisement for neon read: "These GLOWING TUBES never fail to ATTRACT the ATTENTION of a far greater number of people than ordinary signs with their broken line of bulbs."[69]

In the east, other methods of coloring signs at night were retained. For example, exposed incandescent bulbs were either coated with color or surrounded with "color caps," tightly fitted surrounds of colored glass. This practice led to the use of "canteen" globes of colored glass, upon which names or symbols could be stenciled. Globes with letters on them could be grouped to spell out a word, but usually single globes were hung over a lodge hall entrance or a shop door. They represented a token use of electricity in signs. Another technique was to reflect light off of a brightly colored and highly reflective panel. This involved reflecting signs, usually of porcelain enameled metal, floodlit by concealed incandescent filament lamps (fig. 10.6). These signs were commonly found above the doorways of retail stores and were hung from posts in front of gasoline stations. Many national companies, Coca Cola for example, made standardized reflecting signs available to businesses at low cost. The brightly stamped company logo was joined by the name of the establishment, be it a drug store, restaurant, or other business.

Also popular among storefront signs during the years 1910–30 were boxed sign displays, called shadow-graph or shadow-box signs. Light was directed by reflectors through glass surface lenses set in metal panels. Beginning in the 1930s, more and more signs directed light through large panels of acrylic plastic that, indeed, sealed the entire sign face—making

America's Fastest Growing Electric Sign Manufacturing Concern---Selling at Unheard Of Low Prices through a Ford Quantity Production Policy

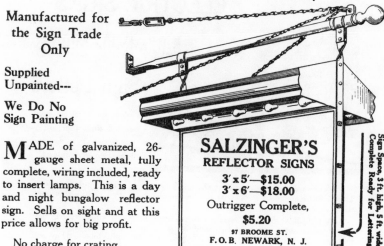

Manufactured for the Sign Trade Only

Supplied Unpainted---

We Do No Sign Painting

MADE of galvanized, 26-gauge sheet metal, fully complete, wiring included, ready to insert lamps. This is a day and night bungalow reflector sign. Sells on sight and at this price allows for big profit.

No charge for crating.

SALZINGER'S
REFLECTOR SIGNS
3′ x 5′—$15.00
3′ x 6′—$18.00
Outrigger Complete,
$5.20
97 BROOME ST.
F. O. B. NEWARK, N. J.

Sign Space, 3 ft. high, 5 ft. wide
Complete Ready for Lettering

Two illustrations below show how these signs look when completed. We do no painting. These signs when built singly cost $40 to $50. Due to our immense production facilities we are able to manufacture them at these remarkably low prices.

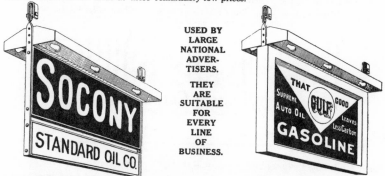

USED BY LARGE NATIONAL ADVER-TISERS.

THEY ARE SUITABLE FOR EVERY LINE OF BUSINESS.

Complete Illustrated Catalog With Prices Mailed Upon Request Sent to Us on Your Business Letterhead

SAMUEL SALZINGER
Manufacturer of Electric Signs for the Trade

93-95-97 BROOME STREET
NEWARK, NEW JERSEY

Now occupying our own three buildings—is ample proof of our growth and success

FIGURE 10.6. Advertisement for the Samuel Salzinger Company. Rising in popularity through the 1930s, reflector signs made of porcelain enamel continued to advertise small retail stores on America's main streets well into the 1960s. From *Signs of the Times* 52 (March 1926): 45.

A NEW DEPARTURE IN ELECTRIC SIGNS—

LU-MI-NUS

Sign Letters

make those **raised gold-beveled signs** that are so easy to read both day and night from any angle.

Your customers realize that the modern sign is the symbol of success. They will be attracted at once by the prosperous, high class air of a sign made with LU-MI-NUS Letters.

A LU-MI-NUS Sign costs less to maintain than any other electric sign. One 25-watt lamp will light two medium sized letters, one on each side of the sign. Your customers can save 75 per cent. on light bills. They will appreciate this.

The signs are easy to build and easy to clean.

You make the signs—we furnish the letters. You earn profits on both the original sign and on re-orders for letters.

Yes—we have some dealer territory still open. Write us for information in regard to this money-making proposition.

LU-MI-NUS

Sign Letters Corporation

NATHAN HERZOG, President.

General Offices:
431-439 S. Dearborn St., Chicago, Ill.

Western Headquarters:
165 Eddy St., San Francisco, Calif.

FIGURE 10.7. Advertisement for the Lu-mi-nus Sign Letter Company. Backlit signs came to the fore beginning in the 1930s, and soon entire signs were lit from behind, not just the exposed lettering. From *Signs of the Times* (Aug. 1919): 49.

the entire panel luminous, not just the lettering. The lamps of enclosed signs were not exposed directly to view; instead, light was diffused through a colored translucent material. These signs produced less glare, and therefore they appeared more tasteful or refined.

At first only lettering was backlit, words silhouetted against the dark background of a supporting metal panel. Either entire letters molded in glass were lit or letters were formed in outline with small round glass

5227-CURTIS STREET AT NIGHT. DENVER.

FIGURE 10.8. Curtis Street in Denver, Colorado, circa 1925. The postcard's caption reads: "Street lighting by electricity has reached a high state of development in Denver. Not only does the city spend a large sum of money annually for its street lighting, but the merchants have acquired the 'electric sign habit' and they vie with one another in producing sign creations that are dazzling in the extreme."

prisms.[70] Enclosed signs required fewer lamps to operate and thus consumed less electricity, and enclosed incandescent bulbs were less exposed to the weather, deteriorated less rapidly, and required less maintenance. The movie theaters of downtown Denver displayed both the older exposed lamp and the newer enclosed lamps in the mid-1920s (fig. 10.8).

FIGURE 10.9. Designs for enclosed signs with backlit plastic panels. Such backlit signs came to dominate retail facades in the 1960s. From G. R. La Wall, "Luminous Wedge," *Magazine of Light* 4 (summer 1935): 25.

In 1923, electrical engineer Matthew Luckiesh estimated that some 56 percent of the electric signs in the United States used exposed lamps, while some 16 percent were enclosed.[71]

Translucent glass was used, enabling sign designers to light entire sign faces from behind. Nonetheless, the backlit panel truly came into its own with development of methyl methacrylate resin plastics and fluorescent tube lamps. An entire sign could be made to glow at night when plastic was lit from behind by the bright new light. New York City's 1939 World of Tomorrow Exposition made wide use of the new technology, displaying the excessive luminance as a significant new departure. Lettering could be silhouetted on acrylic panels that stretched across a building's entire facade, integrating backlighting into a building's architecture.

Luminous fascia displays permitted larger signs than merchants could otherwise afford, given the cheapness of materials and the economies of fluorescent tube operation. As one advocate wrote: "It becomes a distinguishing architectural note capable of individualizing [a merchant's] location; it contributes to the lighting of surroundings which attracts attention to the location generally."[72] Such signs epitomized the modern ideal. Translucent plastic signs with acrylic letters, produced by injection molding processes, came to dominate retail sign advertising in the 1960s. The trade name *Plexiglas* found such widespread use that it become a descriptor for a generic product.

The impulse toward modernism also produced so-called edge-lighting, popular immediately before and after World War II. Letters with small lamps concealed within them were raised above flat background surfaces. At night, when such lettering was illuminated from behind it appeared to "float." Widely used at Chicago's 1933 Century of Progress Exposition and again at New York City's 1939 World of Tomorrow Exposition, such signs lent themselves to buildings designed in streamline modern style. Letters in many such signs were not only backlit but also reinforced with recessed neon tubing. In the dark, the neon reflected from within the letters, making them appear, in turn, to float in space. The open silhouette sign was especially popular with several chain clothing and shoe retailers that used it to convey a sense of design sophistication nationwide. With these signs, the "Great White Way" ideal gave way to subdued, quite unspectacular nighttime visual effects.

Beginning in the 1870s, billboards were standardized in the United States through the International Bill Poster's Association of North America. In 1912, the twenty-four-sheet paper poster board was made standard (with dimensions of eight feet eight inches high by nineteen feet six inches long), although eight-, twelve-, and sixteen-sheet signs remained in demand.[73] Such poster boards in big city downtown locations were frequently floodlit at night (fig. 10.10). In addition, customized painted wall signs and painted billboards were also popular, varying in size according to available space, height above street, and viewing distance. In Detroit's Cadillac Square in 1923, signs were viewed by an estimated 1.5 million people each day (778,700 pedestrians, 610,000 streetcar passengers, and 142,7000 motorists).[74]

Projector signs—using light projected by electric arc lamps or incandescent tungsten filament lamps—never really caught on. Yet it was an idea that kept reappearing decade after decade. In 1892, a large "magic lantern" was installed atop Joseph Pulitzer's World Building, then the tallest structure in New York City. With an eight-inch lens and some 1.5

FIGURE 10.10. Advertising illustration for the General Outdoor Advertising Company. Billboards, no matter how brightly they were floodlit, could not shine in the night like electric signs. From *Poster* 17 (March 1926): 31.

million candlepower, it was intended to project light through stenciled plates of words and picture outlines onto low cloud cover and, on cloudless nights, onto the blank walls of nearby buildings, offering "display cards on the sky."[75] The passing fad grew out of the city's various newspapers' reporting of election-night returns. More practical, but no more popular, were the small lanterns intended to project advertising slides downward onto sidewalks. Through the 1920s, projected slide advertisements would be used primarily in motion picture houses, projected on screens between feature movies.

Signs were made to attract attention, and signs illuminated in the night were, perhaps, the most attracting of all. Critics—advocates of the "City Beautiful," for example—promoted American landscapes free of commercial signs. Reformer Charles Mulford Robinson wrote glowingly of cities devoid of outdoor advertising: "A commercial street, lined with imposing architecture, the facades unmarred by lettering, no screaming signs, by day no glaring colours, by night, no flashing advertisement, all dignity, repose, and self-contained placidity—this would possibly be the vision of the city ideally fair and stately."[76] Most Americans did not necessarily want "all dignity, repose, and self-contained placidity," however. They tended to prefer what another reformer called the "vulgar intrusion on intellectual calm."[77] Signs, electric signs especially, brought much visual excitement to city streets. And they played a vital role in driving mass consumption, the engine of the new American economy. Electric signs brought a sense of playfulness to the American night, not only for pedestrians but more importantly for motorists. Illuminated signs helped make the night visually intelligible and exciting for those who viewed it from behind a windshield.

LIGHTS EMERGED ON BROADWAY to advertise the "good life" for Americans. With spectacular lighting displays, the creation of want and desire was made into a lavish entertainment. It was a blatant commercial manipulation of visual sensitivities, reminiscent of a carnival in its execution. This atmosphere evolved for a number of reasons, all reinforcing of one another. Theater developers sought less expensive real estate up Broadway, bringing with them a concentration of bright marquees. The substantive pedestrian traffic generated by mass transit made Times Square the ideal focus for advertisers wanting to maximize visibility. The image of an entertainment zone of theaters, roof gardens, restaurants, and hotels was further reinforced by the brightness and the color of the signs themselves. The Great White Way was the result of a profound symbiosis.

So strong was the image of nighttime Broadway that even after forces sustaining it had waned—as prohibition closed down the roof gardens and the restaurants, as motion pictures displaced the legitimate theaters, as automobiles replaced mass transit, and as unsavory forms of escape and recreation gained footholds—the Great White Way persisted as an icon of New York City and even the nation. This symbolic center of the nation's most important metropolis was too important to let go. Light filled the sky as sign, signing in the process a profoundly durable sense of place. The Great White Way stimulated imitators elsewhere and greatly popularized the use of electric signs in America. Broadway and Times Square became a kind of nighttime ideal against which communities across the nation measured themselves.

Through the use of electric sign spectaculars, the business of selling was made into an extravaganza with allusions to the world's fair midways and amusement parks. Electric sign art recreated the sense of festival that the use of concentrated bright lights in cities had first conveyed. Brightly colored electric signs, even the most modest, communicated to Americans the promise of participation in a new world of consumption. This promise was not made solely in spectacular sign displays downtown, but, more importantly, it was also made by the smaller merchants who added signs to their neighborhood stores. The whole symbolized a progressiveness and connection to the newly emergent economic order built on widespread access to material goods and services. Electric sign art ultimately introduced a new design vocabulary into the configuring of retail store facades, both on traditional main streets and along the new highway-oriented commercial strips. Electric sign art would come to substantially impact the architecture of the automobile-oriented roadside.

CHAPTER ELEVEN

Lighting America's Main Streets and Commercial Strips

Illuminated storefronts, combined with street lighting and electric signs, made America's main streets exciting at night. In retail districts, large and small, Americans found the necessities of life made available through a rapidly maturing commercial economy. In downtown areas, a kind of equality was demonstrated similar to that in the nation's amusement parks and other entertainment zones. Indeed, shopping was made into a kind of entertainment in which everyone was invited to spend—or contemplate spending—money. Americans in the nineteenth century went from relying on the barter of goods that were largely hand-produced in local households and workshops to choosing from any number of factory-made products marketed nationwide in stores. Corporate enterprises mass-marketed not only material goods but also an increasing array of specialized services. Bright light at night proved an effective salesman.

A new cultural imperative was born, at once materialistic and liberalizing. "Qualities once seen as subversive and immoral and as existing on the margins of American culture," historian William Leach wrote, "gradually moved to the heart of that culture—carnival color and light, wishing, desiring, dreaming, spending and speculation, theatricality, luxury, and unmitigated pursuit of personal pleasure and gain."[1] Of course, that was the lesson that the world's fair midways taught, as did Coney Island and Times Square. And that message was repeated on Friday or Saturday night along America's main streets. There, life became "consumerist" in nature, lacking only the exaggerated sense of whimsy and fantasy that the entertainment zones of big cities possessed.

The automobile took people off the streets as pedestrians and reinvented them as motorists. At first, automobiles reinforced America's localized small town main streets, although higher speeds and the related increased travel range of motor vehicles also opened up new distant shopping opportunities. Small town customers could drive to cities and patronize larger stores there that passed on lower prices through volume sales. Accordingly, main streets in the smallest places withered as the provisioning of goods and services shifted up what geographers came to call the "central place hierarchy." The automobile also changed business patterns within larger towns and cities. Commuters turned away from streetcars in favor of private automobiles, diminishing the viability of commercial thoroughfares oriented to streetcar stops. New automobile-convenient commercial strips began to grow out beyond the edges of towns and cities.

MAIN STREET AS THE "WHITE WAY"

The principal retail district was usually the first part of a city systematically illuminated by street lights. Likewise, "Main Street" was usually the first street to have its lights upgraded when new lamp technologies became available. This was especially the case with electric lights, when tungsten filament incandescent lamps replaced arc lamps in street light luminaires, for example. Communities treated electrification "as a form of conspicuous consumption that said, 'we are progressive, and growing.'"[2] Nothing made a city or town appear more glamorous than "white way" lighting, with globes clustered atop ornamental lamp posts, themselves systematically arranged along streets as much for display as for illumination. Certainly, "white way" lighting made community elites happy. Merchants hoped to expand store hours and increase sales, property owners hoped to charge higher rents and enjoy, accordingly, inflated real estate values, and bankers hoped to expand levels of economic activity.

The "white ways" amplified the vista of the street at night. The bright globes along street margins shined as linear arrays of light while, at the same time, illuminating the facades of buildings that edged the street.

Light edited as it focused, reducing the visual chaos of daytime into a nighttime order.[3] The glare of open arc lamps tended to inject a coldness onto streets, a sense of distance that contrasted sharply with the warmth and sense of intimacy of gas lamps.[4] But enclosed arc lamps and, especially, tungsten filament incandescent lamps, when shielded in globes, introduced a more colorful ambient glow while retaining brilliancy. The light reached out with a magnetic influence, according to one early advocate. "The merchant who appeals to the sense of light in America is clever and on the right track. Babies instinctively look up at a light and reach for anything bright and shiny. Grown people retain something of the child's fear of the dark and are attracted by bright light and 'Great White Ways.' Amusement parks and theaters have for years cashed in on the irresistible attractions people seem to have towards light."[5]

The first ornamental "white way" was installed in 1905 on Broadway Avenue in downtown Los Angeles.[6] According to Walter D'Arcy Ryan, director of General Electric's Illuminating Engineering Laboratory, the purpose of the system was manifold. It was intended, among other things, to provide "cosmopolitan atmosphere and dignified aesthetic effects," minimize glare, cast a uniform illumination flattering to facial features, and outline buildings against a darkened sky.[7] Some 135 posts were equipped with six small glass globes on pendants, each containing a sixteen-candlepower lamp, and one larger globe atop with a thirty-two-candlepower lamp. "With the touch of a magic wand and in a garden of rare color," gushed the *Los Angeles Times,* "Broadway burst into bloom last evening."[8] "White way" lighting fit into "City Beautiful" planning, as it was seen to foster civic pride by making a city more attractive at night. Often it proved a catalyst to other civic improvements.

Broadway Avenue in Los Angeles glowed after dark, the globes illuminating storefronts and outlining upper windows and other recesses in varying shadows. Building cornices, only five and six stories above the sidewalks, were silhouetted against the sky. For pedestrians passing by and for those in the open cars, downtown was made a shimmering background for shopping and other errands. The street appeared extenuated through the regular rhythm of the lampposts and luminaires. At the same time, fixtures glowing opposite one another tied both sides of the street together. As visual display, nighttime on Los Angeles's Broadway was more integrated as a visual set piece. It was not electric signs but street-lights that dominated.

Other cities followed Los Angeles's lead. In Atlanta in 1910, the *Atlanta Constitution* advised readers that the new "white way" lighting on Peachtree Street would be "effective, but not glaring," it would light all the buildings from "top to bottom," and it would "cast over the entire street a radiance which will be soft, brilliant and without casting a single shadow."[9] In Indianapolis, several hundred lampposts, each equipped with five tungsten filament bulbs enclosed in diffusing globes, were installed in 1912. The lights lit the entire downtown region as opposed to a single street.[10] After 1915, the clustering of round globes was supplanted by other luminaire types placed on posts singly, in pairs, or in triplicate. Vase-shaped fixtures with refracting lenses deflected light away from building facades and directed it toward the street. Thus the "white way" rubric came to signify a wide variety of ornamental luminaire configurations.

Critics of the original "white way" lamps pointed to the wasted light sent skyward and criticized the glare that reflected off store facades, diminishing the effectiveness of show windows and electric signs. The excess brilliance from street lighting could "render useless" the merchant's advertising dollar and reduce streets to "deadly monotony"; cities could be "robbed of their individuality at night."[11] In 1915 in Cleveland, posts with single luminaires equipped with strong refracting lenses were installed to deflect "into useful zones the light which would otherwise escape at angles where it would be wasted, or where the glare would cause discomfort."[12]

Business associations initiated and, indeed, subsidized most "white way" developments. "To attract the public to his establishment is the big problem of the retail merchant," wrote Leo Pfeifer in a 1916 article in the trade journal *Signs of the Times.*[13] The logical first step was to attract the public to the street. In St. Louis it was the Civic League's Committee on Street Lighting that decried the poorly lit thoroughfares downtown. The Down Town Lighting Association was organized in 1908, comprised of interested property owners and tenants, and within the year 400 magnetite arc lamps of "artistic standards" had been erected.[14] In Salt Lake City, as in St. Louis, a special tax on abutting property was used to fund lighting

Don't Fail to attend the Meeting

on the subject of the

"GREAT WHITE WAY"

to be held at the Hall of the Merchants and Manufacturers Association, Corner Broadway and Mason Streets on

Tuesday Evening, Oct. 19th

at 8:00 o'clock

It will be the beginning of a Larger Milwaukee

ADMISSION FREE   EVERYBODY WELCOME

LADIES RESPECTFULLY INVITED

DON'T MISS IT

Compliments of the
Chicago-Milwaukee Road & Realty Company
707 GERMANIA BLDG.   MILWAUKEE, WIS.
TELEPHONE GRAND 721

FIGURE 11.1. Postcard promoting a new "Great White Way" in Milwaukee, Wisconsin, circa 1915.

projects. The city took ownership of light fixtures, the Utah Power and Light Company provided electricity and maintenance under contract, and the Commercial Club paid for installation. Luminous 1,500-candlepower arc lamps, placed under vase-shaped luminaires, were clustered atop the power poles of the streetcar company, three to a pole.[15]

Large cities upgraded downtown lighting as older systems proved inadequate for increased motor traffic. Los Angeles replaced the lamps on Broadway Avenue in 1920 to create the "Path of Roses." The lampposts featured flower ornamentation topped by two luminaires with tungsten filament lamps, all subsidized by the Broadway Association. The "Path of Roses" followed San Francisco's 1916 installation of a "Path of Gold" on that city's Market Street.[16] The merchants of Chicago's State Street paid for the relighting of their avenue in an attempt to make it the "brightest outdoor mile ever made," with illumination levels raised to that "used for indoor reading."[17] President Calvin Coolidge threw a switch in the White House to inaugurate the new "Daylight Way" in Chicago, and Thomas Edison sent a congratulatory telegram. Of course, the lamp manufacturers aided and abetted such street lighting initiatives, seeking, as always, to

Make the business district attractive

¶ Good illumination in the streets is naturally followed by increased business.

¶ In a city, with a brightly lighted business section, going "down town" at night is a pleasure which soon becomes a habit.

¶ Out-of-town trade is also attracted and the business streets are enlivened and made more prosperous.

¶ Get the business men and merchants in your town together and go into the matter of installing Ornamental Luminous Arc Lamps, as many other progressive cities are now doing.

¶ This efficient system of lighting does away with unsightly poles and indifferent illumination.

¶ The Ornamental Luminous Arc Lamp gives a beautiful, white, well-diffused and brilliant, though soft illumination.

¶ The artistic poles lend themselves to various designs —making the streets attractive by day as well as night.

¶ Send for Bulletin No. 4955 which gives details of this most up-to-date system of Ornamental Street Lighting.

¶ Special information will be gladly given on the "White Way" type of lamp for your city.

General Electric Company

General Office:
Schenectady, N. Y.

Sales Offices
in All Large Cities

FIGURE 11.2. Advertisement for General Electric's luminous arc lamps. Light expanded the urban market by extending business hours. Several nights a week, downtown stores remained open in most cities and towns through the early evening hours. From *American City* 9 (Dec. 1913): 44.

firmly tie artificial illumination to progressive business practice. One General Electric advertisement read: "Get the business men and merchants in your town together and go into the matter of installing ornamental luminous arc lamps, as many other progressive cities are now doing" (fig. 11.2).[18]

Though big cities adopted "white way" lighting first, the idea spread to small towns almost immediately. Emulating Milwaukee, Fond du Lac in Wisconsin adopted ornamental clustered globes in 1912. Funded by the Business Men's Association, fifty-eight posts with tungsten filament bulbs were erected along the town's main street. An upper lamp burned throughout the night at town expense, while the lower four lamps were lit only in the early evening hours, at merchant expense. An article in the *American City* reported the transformation of the city's principal business street from "a congested thoroughfare cluttered up with poles and networks of wires into a handsomely illuminated boulevard." The brightening of Fond du Lac's downtown district convinced several merchants to install new, "modern" storefronts. "A feeling of optimism and a spirit of cooperation" reportedly overspread the town.[19]

Civic improvement through street illumination also was promoted in the electrical engineering journals. In 1896, Fred De Land wrote of a Michigan town, "Now the hanging of those big lamps in our streets has indirectly wrought as marvelous a change as often follows the placing of a new and attractive carpet in the reception-room of the old homestead. Gradually the old style furniture is replaced by the modern and more attractive until finally the room is modernized throughout. . . . So it was in our town." He continued: "The arc lights gave us a taste of progressiveness and awoke a desire for other improvements." The town's business streets were paved with brick, and several frame stores ("that had been a fire menace for years") were torn down, making way for a modern office building erected on their site.[20]

Such homely endorsements brought nothing but ridicule from Sinclair Lewis, who debunked, generally, American provincialism. "Then glory of glories," he wrote of his fictional Gopher Prairie in his novel *Main Street,* "the town put in a White Way." "White Ways were in fashion in the Middlewest," he continued. "They were composed of ornamental posts with clusters of high-powered electric lights along two or three blocks on Main Street. The *Dauntless* confessed: 'White Way Is Installed—Town Lit Up Like Broadway—Speech by Hon. James Blauser—Come On You Twin Cities—Our Hat Is In The Ring.'"[21] Nonetheless, such pronouncements, built around the lighting of America's small towns, seemed to "actualize dreams of greatness," commented historian David Nye. Lighting was more than a mere functional necessity, it was "a glamorous symbol of progress."[22]

DISPLAY WINDOWS

The display window first appeared in America's eastern seaboard cities in the 1830s. Through the early nineteenth century, retail shops were usually located in first-floor front rooms of townhouses, with shopkeepers and their families residing in rooms above and behind. As New York City, Philadelphia, and other urban centers grew, older townhouses were converted completely to commercial spaces and the more affluent business families moved into new, exclusively residential neighborhoods. With increasing frequency, shops came to occupy new and larger structures built especially for business but still similar in outline and floor plan to the old row or terrace houses. The business "block" evolved. Over front doors hung signs, sometimes with lettering but just as often with mere emblems to signify businesses within. With time, front windows developed into display areas, which, like the signs, invited illumination during evening hours.[23]

The shop window became a kind of glassed-in stage, especially after plate-glass became available around 1850. Customers on the sidewalk could see not only merchandise "in the window" but also the entire shop interior, no longer obscured by the mullions of traditional window panes. The showcase in a store window was an invitation to enter. At first, plate-glass was imported from France, but then an American industry organized and, by 1915, Americans were consuming one-half of the plate-glass manufactured worldwide.[24] Gas lamps required a merchant's attendance, since there was fire danger and, when lamps were left burning through the night, undue moisture and soot could build up inside a store. Electric lamps, on the other hand, could shine unattended until morning—beneficial not only for their advertising value but also for security.

The application of electric lighting to display windows was at first a matter of attracting attention, rendering a store vivid all night long in comparison to darkened storefronts nearby. Neighboring merchants also quickly adopted electricity in self defense; when a new form of electric light appeared in an American city, it was sure to spread from store to store and from street to street.[25] By 1891, however, trade journal editors were cautioning: "Proprietors of stores are beginning to recognize that abundance of light alone . . . is not everything, and that to attract the public something more is necessary." "That something," one editor continued, "is a combination of the judicious disposition of light, its fine gradations, its covering, its accompanying fittings, and its tone or coloring."[26]

Fully lit store windows provided much of the nighttime illumination for sidewalks and streets before "white way" lighting was introduced. Store windows were, of course, major attractions not only for their brilliance but also for the sophistication of their displays. A new profession emerged, represented by the National Association of Window Trimmers, founded in 1898. The arrangement of merchandise, "show cards" describing items, background motifs, and lighting effects received increasingly sophisticated attention.

"Pictures are better than word descriptions," wrote S. L. Ruzow of the Sarnoff-Irving Hat Store chain, "but the actual goods itself [sic], displayed in an attractive manner, invariably will create a stronger incentive to purchase."[27] The *Journal of Electricity* advocated illuminated displays for several reasons. Electric light added "attractiveness and value" to merchandise displayed. It permitted closer and more accurate inspection of goods, thereby reducing the number of exchanges. It increased the rental value of stores located in the middle of blocks by making them as visible as higher-rent corner locations. It created a "silent salesman" after closing hours to serve the window shoppers.[28] Display window illumination also could be the stuff of poetry. James Stuart Montgomery penned his "In the Windows," a poem expressing the post–World War I desire to go "back to normalcy." It reads, in part:

> *Let's go down and see the glow*
> *Of windows, row on row;*
> *Poppied shelves of elfin wares,*
> *Silver citrons, golden pears,*
> *Raisins from Avoca vale*
> *And pomegranates are on sale.*
> *Let's go down and cast away*
> *Shadows grim and shadows gray.*[29]

FIGURE 11.3. Advertisement for the Pittsburgh Reflector Company. Through the 1920s, display windows were crowded with merchandise. Brightly lit, they offered an almost encyclopedic understanding of what merchandise awaited inside. From *Chain Store Age* 5 (March 1929): 73.

Through the 1920s, most display windows were packed with merchandise. The trend was a carry-through of the world's fair exhibits, where, inside the massive exhibition halls, display cabinets were crowded with curiosities. The display window was a three-dimensional catalog of the merchandise to be found inside. Light from concealed sources was carefully diffused, often during both night and day. Daylight illumination was calculated to eliminate street reflections on window glass. In clothing stores, lifelike wax mannequins appeared about 1910. As large objects, they required more space and the elimination of display window clutter in order to show effectively. Window trimmers began to display fashions on mannequins against backgrounds that implied a sense of place—to develop a sense of social context conducive to purchase. Display became more a medium of fantasy, with lighting playing a critical role in contriving theatrical contexts.

Beginning in the 1930s, especially in large downtown department stores, display windows served more to "set mood" and less to display merchandise. Windows were intended to attract customers inside onto sales floors laden with merchandise while at the same time reinforcing if not raising customer lifestyle expectations. It was not just the merchandise that was offered; the pleasure of buying that merchandise was also included in the price. Purchasing an object in an atmosphere of affluence and well being was also what counted. Department store windows were changed to reflect the seasons. At Lord and Taylor in New York City, the 1951 fall windows emphasized resort wear, anticipating vacations in the Caribbean. In one display, brilliant Haitian settings provided a background for black beach clothes. A sign read: "New thoughts for tropical climes where magic is black."[30] For the Christmas season of 1952, one Bonwit Teller window depicted the "Man-in-the-Moon" gazing upon the "Queen of the Night." "Clouds of the airy white spun glass which covered the floor of the window and the silver-blue lighting further sustained the mood of the setting." Female mannequins wearing negligees stood among small blue and white Christmas trees, bearing gift suggestions.[31]

FIGURE 11.4. Bleecker Street at night in Utica, New York, circa 1940. The eye is carried along the street by the bright street lights, illuminated display windows, and lit advertising signs. Courtesy of Lake County (Ill.) Museum, Curt Teich Postcard Archives.

THE STOREFRONT AS SIGN

The electric sign and the illuminated store window worked together on the well-lit retail street. At first, electric signs were usually located over doorways, but later they appeared over display windows as well. The more aggressive retailers placed additional signs on upper stories and rooftops. "Pretty soon the town is ablaze, and it is noticed that the merchants develop civic pride enthusiasm, which is passed through their clerks and through the newspapers to the public," insisted the *Signs of the Times*. "The show windows begin to develop a stronger attraction than the fireside, and the municipality graduates from the town class into the city class," the article continued.[32] English author J. B. Priestley marveled at the effect of the bright signs even in the isolated small towns of the Southwest. "And every village and tiny town, with its neon signs, looks to a European like a bit of a city that has just detached itself. There are the drug stores, the eating-houses, the hotels, all blazing with colored lights. They may be a hundred miles from anywhere, deep in the desert, but you would imagine that somebody had contrived to pick out a block from Sunset Boulevard, Los Angeles, and drop it down there." "At night," he mused, "you travel either in the darkness of Siberia, or through what appears to be the main street of a carnival town."[33]

Vertical signs hung over sidewalks and horizontal fascia signs set flush with building facades added to the carnivalesque glimmer, sending a colorful sheen over the street and reflecting vividly in the glass and metal of parked cars. Many signs blinked, sending letters and words through ritual motions and adding to the sense of gaiety. Even when the stores were closed, a sense of visual excitement was sustained.

In the 1930s, as economic depression impacted America's main streets, storefront signs and display windows began to merge. Tacked onto the National Housing Act of 1935 were provisions guaranteeing remodeling loans for storefront refurbishing, and a campaign to "modernize" America's main streets was launched as part of the Roosevelt Administration's "pump priming" tactic of using the construction industry to jumpstart the American economy. The tenets of architectural modernism were emphasized, encouraging smooth, unornamented, open facades clad in plate-glass, structural glass, porcelain enamel, stainless steel, and aluminum, which were all either translucent or highly reflective of light.[34] Plate-glass allowed storefronts to be opened up at ground level by eliminating traditional display windows. The entire store could be fully exposed, making its interior into a giant display window, especially under the bright and even illumination of fluorescent tubes. Setback entrances with rounded, "bent glass" corners invited easy entrance. Up above, backlit signs, sometimes covering the remainder of a facade, continued the nighttime glow.

The rise of interior fluorescent lighting and air conditioning made windows redundant in upper stories. Those could be covered by screens or facings, making an entire storefront into a giant sign to be backlit, floodlit, or otherwise illuminated. The trend toward "open" storefronts accelerated after World War II and came to dominate not only on reconfigured traditional main streets but also, more importantly, the new shopping districts out along suburban commercial strips. Signs, windows, and store interiors were now designed as a single unit, as nighttime illumination was made a point of departure for architectural planning. Particularly in the

FIGURE 11.5. Advertisement for the Curtis Lighting Company. By 1940, fluorescent lighting literally had brought store interiors to daylight levels of brightness. The new lighting facilitated "self-service" merchandising, with goods displayed on open racks and shelves rather than behind service counters. From *Chain Store Age* 17 (Nov. 1941): 27.

new suburban shopping centers, business hours were pushed to begin later in the morning and extend further into the evening. The illuminated storefront as sign became crucial to retail success there.

THE STRIP

America's traditional main street retail buildings formed around a standard template—the storefront that was flush with the sidewalk and street and had an entrance and display windows set in a flat wall, usually containing upper-story windows and capped by an ornamental cornice. It was a three-part formula into which signs usually intruded as applique. As architectural historian Chester Liebs wrote: "Signs covered windows and spandrels, blanketed exposed walls, projected out from the building facades, and jutted up from roofs high above the cornice line."[35] A tug-of-war was initiated between traditional building styles and bold commercial signs; sign and storefront eventually merged on America's downtown streets, as described, through the mechanism of the display window as open storefront.[36]

Automobile-oriented roadside strips emerged in the 1930s, blossoming after World War II out along open highways beyond the edges of towns and cities. There land was cheap, and large parcels could be assembled to accommodate spread-out building footprints as well as extensive driveways and parking lots. Commercial areas that were distributed along the axes of streetcar lines were organized in more closely packed storefronts oriented to pedestrian traffic, just as in downtown business districts. Such clustering facilitated multistop shopping. Out along peripheral highways, however, businesses were spread out, and single-destination shopping prevailed. Businesses at the edge of cities could intercept out-of-town and suburban customers who appreciated not having to penetrate congested city centers, as well as tourists or other travelers who sought services but also hoped to avoid downtown traffic. Along the new roadsides, businesses tended to be in single-story buildings, as automobile traffic was insufficient to support upstairs offices and the sites were too far removed from other urban amenities to support rental apartments. Set back from highways by driveways and parking lots, the low-profile buildings were

FIGURE 11.6. Dan's Drive in at Columbus, Ohio, 1995. A giant roadside sign was essential to attracting motorists to businesses housed in low buildings and positioned well back from public rights-of-way, as along suburban commercial strips.

very much dependent upon large signs out front in order to attract customer attention day and night (fig. 11.6).

Only at various shopping centers were stores clustered to reinforce one another in the traditional business way. Yet there the template of the isolated business was repeated for the collective good of grouped tenants— to support a giant sign at the street, a sprawling parking lot, and low-rise building complex. True, the huge shopping malls later turned inward to replicate main street in air-conditioned and brightly lit atriums. Through floodlighting and the positioning of signs on the taller anchoring department stores, however, shopping malls were rendered as identifiable as gasoline stations, fast food restaurants, and other strip businesses. They were simply giant signlike structures.

As Liebs asserted, architects had traditionally strove to express character in buildings without resorting to advertising signs. Rarely were commercial buildings designed with the inevitable outdoor signs in mind. Consequently, signs were often "plastered in every conceivable noticeable location, causing eyesores and effacing all," according to journalist P. Schuyler Van Bloem in 1929.[37] But, under the hand of the modern architect, Van Bloem continued, electrical signs would "find their proper place and form and, instead of monstrosities from the shops of graduated carpenters, electricians and tinsmiths, electric signs [would] become beautiful, harmoniously ornamented panels worked into the worthy architecture symbolic of America." Indeed, the electric sign did come to substantially dominate roadside architecture. However, it would be a new kind of building tradition, reflective of the new automobile culture. Ironically, this new architecture would not be the work of architects after all. It would be a new vernacular form put up by owners with limited experience and taste, following the guidance, if guided at all, of electrical contractors.[38]

Substantial impulse shopping was conducted along the roadside, as people moving at thirty or forty miles per hour decided at a glance where to stop.[39] Low-profile buildings needed to be complemented visually by large signs set out at road margins, marking driveway entrances. Pylon signs soared up thirty, fifty, or seventy feet, with messages boldly lettered in neon tubing or in incandescent bulbs for night viewing. Signs were attached to buildings as well, usually above open or "visual" fronts completely glazed and open to view. Glass was often canted inward to reduce reflections, and roofs were often cantilevered up and out to increase the amount of glass surface.

Builders embraced exaggerated structural components.[40] "Butterfly" and "hyperbolic parabolic" roofs not only enhanced roadside structures

FIGURE 11.7. Country Kitchen restaurant at Wisconsin Dells, Wisconsin, circa 1960. Advertised as "Wisconsin's most unique," this restaurant was but an early introduction of roadside "exaggerated modern." The building sits at the roadside glowing like a giant television set, the interior totally open to view. Courtesy of Trumble Photography.

but also provided platforms for signs. Towers were used as sign platforms, further integrating sign and building. The use of what writer Tom Wolfe called "exaggerated modern" forcibly transformed buildings into signs. Wolfe wrote of Los Angeles and the endless scorched boulevards lined with one-story stores, shops, bowling alleys, skating rinks, taco stands, drive-ins, "all of them shaped not like rectangles but like trapezoids, from the way the roofs slant up from the back and the plate-glass fronts slant out as if they're going to pitch forward on the sidewalk and throw up."[41] At night, these buildings exhibited their insides like giant television sets.[42]

Thus the facade of the Country Kitchen Restaurant glowed in the night (fig. 11.7). Out front the sign loomed, the restaurant's name vivid in the darkness even at some distance. Yet it was the restaurant interior that spoke the loudest. Opened up to the outside through an expanse of plate-glass, interior furnishings took on a hard-edged look owing to the bright fluorescent ceiling lamps. Counters, tables, and booths were easily discerned from outside, the brightness communicating newness, cleanliness, and safety among other measures of worthiness. Neon and fluorescent tubes and incandescent bulbs combined with glass transparency to advertise nighttime opportunity through clean, simple architectural expression.

Journalist Thomas Hine has used the term *populuxe* (an amalgam of *popular* and *luxury*) to describe architectural styles and other trends of the 1950s. Exaggeration and overindulgence were aspects of a decade during which designers turned out fantasy as product. "Color and styling were applied to objects that had always been viewed as purely practical," he wrote.[43] Architecture took on the look of being in motion, but the roundness of streamlining was out, replaced by the square, or "sheer" look. Landscapes of pop art with "bright colors, bold delineations, and popular symbolism" were created, wrote journalist Alan Hess. "The wild forms, glinting surfaces, overscaled hot dogs, clusters of gigantic billboards, private and public fantasies, and neon districts rivaling the most garish sunsets showed a heedless disregard of good taste as it was conventionally known and practiced."[44]

To achieve maximum visual impact, planners used both pylon signs and signlike buildings, combining them to anchor driveway and parking lot. Signs were molded into shapes far beyond simple geometry—creative signmakers fashioned amoeboid-shaped signs, signs replicating space satellites with bristling antennae, kidney-shaped signs, or signs looking like boomerangs. Signs came to represent the strip as a basic icon. "Whole strings of places competing, clamoring, blasting your eye for attention. The golden arches of McDonald's . . . rising triumphantly beside the Rhode-Island-red and white stripes of Kentucky Fried Chicken. The delicate crown of a Dairy Queen poised regally beside Mobil's spinning red Pegasus. And the car and parts dealers, Midas Mufflers, furniture showrooms, pizza parlors, and chili stands—America," wrote landscape critic Stephen Kurtz.[45]

Holiday Inn's "great sign" (fig. 11.8) reflected the background of the motel chain's founder, Kemmon Wilson. As a building contractor in Memphis, Wilson also had operated several small movie theaters. The Holiday Inn's sign, therefore, was a replication of a movie theater marquee; the sign extravaganzas of America's small town main streets had moved to the nation's commercial strips. The "great sign" became a symbol of reliability, advertising the motel of "no surprises," where expectations of high

FIGURE 11.8. Holiday Inn at Liberal, Kansas, 1983.

standards prevailed. It shouted its message into the night with a sense of carnival good humor. It stood as a symbol of 1950s exuberance, reflecting the decade when corporate America heightened its interest in the American roadside as marketplace.

John B. Jackson called the signs and their buildings "other-directed architecture," a new architectural form. "In all those flamboyant entrances and deliberately bizarre decorative effects, those cheerfully self-assertive masses of color and light and movement that clash so roughly with the old and traditional there are, I believe, certain underlying characteristics which suggest that we are confronted not by a debased and cheapened art, but by a kind of folk-art in Mid-XX Century garb," he wrote.[46] It was architecture largely devoid of the practical, workaday world. Instead, it suggested "pleasure and good times" and "the atmosphere of luxury, gaiety, of the unusual and the unreal." It abandoned the imitation of "the everyday activities of a superior social group"—the traditional arbiters of taste. Now a more affluent working class, with increased leisure time at its disposal, could and did create its own styles of, and settings for, leisure activity.

One of those leisure activities—putt-putt golf—carried allusions to the

FIGURE 11.9. Postcard advertisement for Putt-Putt Golf Courses, Fayetteville, North Carolina, circa 1960. Working-class leisure-time pursuits dominated the new roadside strips both in function and symbolically through pop art. Courtesy of Putt-Putt Golf and Games.

aristocratic game it parodied. But golf's version by the roadside was anything but elite. It was pure entertainment requiring little if any skill. Unpretentious relaxation under bright lights was the attraction of this "working man's" sport. It could be engaged on the spur of the moment, attracting motorists purely on impulse, and it was an inexpensive sport that did not consume substantial time. The fluorescent tube lighting used for putt-putt golf sharpened visual outlines, making textures stand out. Though not always flattering to facial features, the bright fluorescent lighting was, nonetheless, by itself, a novelty worth experiencing, a subtle part of the 1950s roadside allure. Business entrepreneurs and their contractors were provided the mood-setting flashiness that suggested the novel and the exotic. "Neon lights, floodlights, fluorescent lights, spotlights, moving and changing lights of every strength and color—these constitute one of the most original and potentially creative elements in the other-directed style. It would be hard to find a better formula for obliterating the workaday world and substituting that of the holiday than this, nighttime and a garden of moving colored lights," Jackson wrote.[47] Nowhere was this reorientation made more vivid than in Las Vegas.

LAS VEGAS

Should one doubt the significance of the commercial strip as a new urban form, one had only to look down from the comfort of an airliner. At night over the Midwest, one could look down on the lights of a small town—"and out of it, like the tail of a comet, stretches a long sinuous line of lights in every color and intensity, a stream of concentrated, multicolored brilliance, some of it moving, some of it winking and sparkling, and every infinitesimal point of color distinct in the clear night air," Jackson wrote.[48] The most spectacular strip of all—the archetype strip—evolved in Las Vegas. At Las Vegas the strip was raised to a new power. There it provided a template for a new kind of American city, one totally oriented to the roadside.

Las Vegas began its rise in the 1930s with two important events, the legalization of gambling and completion of the nearby Hoover Dam. After World War II, a vigorous entertainment industry developed around legalized gambling, and the city came to use cheap electric light to symbolize itself in sign spectaculars the likes of which even New York City's Broadway had never seen. In its early days as a railroad town, the gambling halls were at first concentrated along Fremont Street, immediately east of the Union Pacific tracks. There, led by the Golden Nugget Casino, an extravagant main street display of neon evolved into "Glitter Gulch." By 1950, two wooden "falsefront" buildings had been joined, linked by modest fascia signs lit by neon tubing and incandescent bulbs. In 1962, a giant sign 120 feet tall was erected, containing roughly 9,000 feet of neon tubing and 17,000 incandescent bulbs.[49] None of the original architecture of the street was visible by the late 1960s, buildings having been completely veneered in signlike facades. By 1980, the Golden Nugget building had been completely wrapped with neon and incandescence (fig. 11.10).

Among the sign companies at work were the Young Electric Company (creator of the Golden Nugget display), Federal Sign Company, Heath and Company, the Larsen Sign Company, and Sign Systems, Inc.[50] Their combined activity totally remade the street as a place. Indeed, Fremont Street was closed to traffic in the 1990s and converted into a pedestrian mall featuring neon light shows overhead. The railroad was relocated to make

FIGURE 11.10. The Golden Nugget and Fremont Street, Las Vegas, circa 1983. The post-card caption reads: "This fabulous 'new look' on Fremont Street in the heart of the world's largest gambling center is a sight to be long remembered. The new electric neon signs on the Horseshoe Club and the Golden Nugget light up the sky with a great blaze of color." Courtesy of Western Resort Publications and Novelty.

room for more casinos, the nucleus of the old main street all but obliter-ated. A place for play was created, one in which ordinary life could be transcended. "A mood of festive unreality must prevail," wrote Stephen Kurtz, "outdoors, night becomes day through extravagant amusement-park forms defined by brilliant lights, while, indoors, complete reliance upon artificial illumination warps the sense of time."[51] Gambling could function like a drug, putting its more dedicated practitioners into states of time suspension with the aid of an illuminated environment.

Las Vegas's vitality spilled out beyond the small downtown area to create a new city along the Los Angeles highway, at the far end of which sat the airport. Increasingly after 1950, a great proportion of visitors to Las Vegas arrived from Los Angeles by automobile and by air. Indeed, Las Vegas was really an extension of Hollywood. As lavish hotels and nightclubs were joined to the casinos in integrated resort complexes, fantasy themes from the movies were adopted, while entertainers for the clubs were drawn from Hollywood's pool of talent. Even before Disneyland and other theme parks were created, such themes played out in Las Vegas. The "Old West" was celebrated at casinos like the Pioneer Club, the Frontier, the Hitching Post, the Gold Dust, and the Golden Nugget. The desert atmosphere was reflected in the Sahara, the Dunes, the Oasis, and the Sands. "Fun" (both family and adult) was showcased at the Showboat, Circus Circus, Mardis Gras, and Moulin Rouge. Entrepreneurs like mobster Bugsy Siegel and movie mogul Howard Hughes helped translate the celluloid glitter of "Tinsel Town" into the giant movie set of Las Vegas.

Whereas the traditional storefront was the template from which Fremont Street evolved as Glitter Gulch, it was the roadside motel that provided the template for "The Strip."[52] Starting with the El Rancho Vegas Motel in 1941, the idea of mixing gambling, overnight accommodation, dining, and entertainment in a single large complex gained momentum. As a creature of the highway, the motel-casino was low in silhouette and pushed back from the road by the open expanse of driveways and parking lots, the typical commercial strip presence. As on every strip, the bold pylon sign at curbside brightly lit at night served to create a sense of place. At first, place distinctiveness, casino to casino, was primarily dependent on the signs. Tom Wolfe wrote: "They tower. They revolve, they oscillate, they soar in shapes before which the existing vocabulary of art history is helpless. I can only attempt to supply names—Boomerang Modern, Palette Curvilinear, Flash Gordon Ming-Alert Spiral, McDonald's Hamburger Parabola, Mint Casino Elliptical, Miami Beach Kidney."[53] Yet each sign offered its information according to a set formula. First the emblem and name, often fused, rose up from the road, setting a particular expectation, followed below by an informational box giving headlines and details in smaller print (fig. 11.11).[54]

As on Fremont Street, casino facades on the strip also were covered by sign spectaculars. In 1958, the Young Electric Company built the Stardust Hotel a sign 217 feet long and twenty-seven feet high displaying planets and comets that pulsed neutrons and radiated cosmic rays (fig. 11.12). The letters of the sign varied from seventeen to twenty-four feet in height and sported white incandescent bulbs outlined by blue neon tubing. The sign used some 2,000 incandescent bulbs and 1,500 feet of neon tubing.[55] The colors used in the new signs suggested the full range of exotica popularized by Hollywood animators. "Such color!" gushed Tom Wolfe, "tangerine, broiling magenta, livid pink, incarnadine, fushia demure, Congo ruby, methyl green, viridine, aquamarine, phenosafranine, incandescent orange, scarlet-fever purple, cyanic blue, tessellated bronze, hospital-fruit-basket orange."[56]

Collectively, the signs of the Las Vegas Strip dwarfed those of Times Square. Alan Hess derided Broadway neon as an "urban retrofit." "It used neon like pen and ink, emphasizing thin, two-dimensional lines."[57] But in Las Vegas the signs were three-dimensional. The giant pylon signs were faced on both sides, thrusting up into the night sky to excite people both coming and going. The giant fascia signs on the casinos themselves made entire buildings into signs that blazed in the night in multiple directions. Indeed, it would be more accurate to say that the signs were turned into architecture. Las Vegas signs, like those at Times Square, were animated through "oscillation, scintillation, chasing, flows and sweeps of light."[58] But in New York City they functioned as mere applique, taller buildings soaring all around to define a larger sense of urban context. In Las Vegas, the signs defined their own context. "At night," observed John Pastier, "the dark void and the bright moving lights of the casinos and their signs transform the amorphous space into a tangible one bounded by brightness rather than surface."[59]

As noted in the introduction, interpreting the Las Vegas Strip as an urban form began in earnest in the 1970s with the now classic *Learning from Las Vegas* by architects Robert Venturi, Denise Scott Brown, and Steven Izenour. The Strip was seen as an architecture of communication

FIGURE 11.11. The Circus Circus Hotel and Casino sign, Las Vegas, 1983. The giant curbside sign, combined with driveway, lit porte cochere, gaudily signed casino building, and adjacent parking lot, was part of a design template replicated up and down the Strip.

FIGURE 11.12. The Stardust Hotel and Casino, Las Vegas, circa 1960. Electric signs dominate the casino building, which is, itself, hardly visible in this nighttime view.

rather than an architecture of space. "Architecture in this landscape becomes symbol in space, rather than form in space," they wrote.[60] The message system of the 1970s strip involved giant pylon signs, building facades as signs, their location vis-à-vis one another, and the furnishings of the street. Streetlights were superfluous as sources of illumination, but the posts and luminaires, along with the measured regularities of driveway entrances and the position of median turning lanes, did set a rhythm that served to tie the Strip together visually. The large casinos were located at intervals, and in between them the highway was lined with lesser signs and small sign-buildings—gasoline stations, fast food restaurants, wedding chapels. Though the whole bordered on visual chaos, a sense of pattern was nonetheless visible at speeds of thirty or forty miles per hour. Chaos was near, but its avoidance spoke decidedly of order. The logic of the strip clicked most powerfully at night, when the signs focused attention and the surrounding darkness edited out the unimportant.

The casinos beckoned to the motorist in a formula repeated constantly over six miles of thoroughfare. First, a giant pylon sign invited a motorist to turn into a casino's driveway. Casino entrances were configured after

1970 with huge porte-cocheres, themselves turned into lavish light displays at night. Nearby, a token parking lot suggested complete ease of access. By the 1980s, however, the Las Vegas Strip had changed substantially. Behind the casinos, now enlarged into their former parking lots, soared high-rise hotel towers. Behind the towers were new parking decks. Rather than a skyline of giant pylon signs, the Strip now presented another skyline of twenty- and thirty-story hotels.

The porte-cochere was a master stroke of architectural innovation. It created for the massive casino complexes a clear sense of entrance, especially at night. There was no question as to a potential guest's destination. The casino's persona as a place of substance was reinforced in bright lights. It was clear that behind the entrance lay the full range of resort services expected. Nowhere, save perhaps under the nation's theater marquees, had the sense of entrance been made so evocative. There was nothing subtle about a casino's attempt to snare customers off the street. A sense of grand arrival was promoted.

The Las Vegas Strip continues to evolve. Today, many of the older casinos are gone, replaced by giant high-rise hotels, and the Strip is much enlarged, with many new giant casino complexes. The new casinos personify the "one-stop" shopping of the automobile strip. The arriving guest will find everything necessary within the pedestrian areas of the single casino—gaming rooms, restaurants, nightclubs, gift shops, swimming pools, exercise rooms, conference facilities. The Las Vegas casino has become a resort "shopping center," and the Strip is essentially a linear array of such centers. Only with some inconvenience does the visitor move on foot between casinos spaced at roughly one-block intervals. Mass transit operates, but its intended purpose is primarily to carry the unskilled domestics and other workers who clean and maintain the casino complexes at low wages.

THE IDEA OF MAIN STREET AS A DOWNTOWN RETAIL CENTER began as a daylight proposition. Yet fully illuminated stores could function as effectively, if not more effectively, after dark. Electric lighting served to highlight merchandise, showing it clearly to the potential customer on the sidewalk. Electric signs attracted attention and, in combination with display windows, came to convert entire building facades, if not entire buildings, into signlike structures. Brilliantly lit, the modern store reached out to customers in their leisure hours in ways not at all unfamiliar to seekers of relaxation and pleasure at amusement parks. Indeed, through artificial illumination shopping was made more than ever a kind of entertainment, a kind of recreation.

The commercial strip arose to compete with main street by adapting two important technological innovations to its purposes. First, America's commercial strips were configured to accommodate the automobile, offering driveways and parking lots. Second, the strip adopted modern lighting, featuring fluorescent tube lights inside that could fully display interiors through plate-glass windows to outside view. Pushed back from the road by parking lots, roadside businesses adopted the giant sign both at curbside and on building facades. Fascia signs and supporting structures merged as architecture. Thus businesses attracted at the street, invited through ease of access, and reinforced through inside exposure. This was the formula that spoke to motorists through amplified use of electric light. Main street could not compete; it could only adjust, remodeling storefronts in efforts to adopt the trappings of the new roadside modernity.

The strip was made a place of entertainment through the use of electric light. Commerce and recreation were linked through use of brightness and color, as developers exploited levels of brightness originally reserved for the entertainment zones of world's fair midways, amusement parks, and Times Square. Taken to its ultimate excess in Las Vegas, the strip defined a new kind of city space, if not a new kind of city built solely around entertainment. Initially, the Las Vegas Strip was like thousands of other peripheral business thoroughfares creeping outward from more congested urban space. In the Nevada desert, however, a strip archetype was born. It generated an energy capable not only of orchestrating the essential strip elements but also of raising them above and beyond all expectations. In Las Vegas, the rise of popular culture declared a complete victory over cultivated tastes.

Perhaps the strip represents a new kind of American main street. It is a place gauged to a public moving by automobile rather than on foot. It is a place of exaggerated exuberance played out in sign, building as sign, and sign as building. Its exclamation point is the excessive use of electric light. Americans have wholeheartedly embraced the strip as an organizing device with which to reconfigure towns and cities. After World War II, peripheral roadside strips came to provide the essential frameworks upon which new housing was oriented, subdivision by subdivision. The strips of the 1940s and 1950s are now surrounded by urban development. These strips are central in city space, and, to a considerable extent, their centrality has come to define the contemporary city.

Epilogue

By 1970, the influence of the automobile on nighttime lighting was felt in its entirety; the early attempts at saturating the urban night with light had been improved upon, yielding substantial results. Cities were lit primarily to facilitate the movement of motor vehicles. Downtown in metropolitan areas, the visual effects of other lighting—the informative and celebratory uses of light—literally paled in comparison. A bland homogeneity had come to replace the lighting diversity that earlier had seemed fundamental to modern urban nighttime character. The same was true out along the new commercial strips to which substantial retail and other commercial activity had shifted. Only there, new commercial investment served to generate some respite. Large curbside signs and signlike buildings both gaudily lit the night, offering some counterpoint to modern street lighting, even though this departure seemed, to many observers, vulgar and chaotic. Lit signs were intended not to stand out in dark shadow so much as to stand out in the spillover haze of bright streetlights. Lighting the urban night had been made fully responsive to automobility.

Were we able to observe firsthand the progression of nighttime lighting in American cities, we would be struck by the differences in artificial illumination evident decade to decade. For any point in time, we would likely be surprised by one or another aspect of what we saw (or did not see), like Jack Finney's time travelers in New York City's Broadway of the 1880s. I am sure that we would find much to appreciate. Not everything about previous technology, I suspect, would loom as problematical. Likewise, if we observed the lighting improvements from the nineteenth century onward, not every change witnessed would seem completely progressive. The 1880s and 1890s were decades of decided experimentation, as electric arc and electrical incandescent filament lamps were applied to street lighting. Gas mantle lamps, perfected to meet the new competition, appeared also, adding variety as well as brilliance to public rights-of-way. Between 1900 and 1920, electric incandescence was brought both to outdoor advertising and to the lighting of store display windows. Incandescence was also turned to festive uses of various kinds, encouraging the celebration of cities as places. Celebratory light—for example, the floodlighting of landmark buildings—held its own in the urban night as long as nighttime was still dominated by dark shadow. The 1920s saw street illumination exaggerated in order to accommodate motor vehicles, although neon was soon used on the roadside and partially offset street lighting's overwhelming visual effect. But in traditional big city downtown areas and even on small town main streets, a spillover haze of reflected light came to fill the night sky.

The Depression years of the 1930s and the wartime years that followed saw little fundamental change in how American cities were lit. At the world's fairs, however, new technologies were demonstrated year after year. Americans became accustomed to the very different qualities of light cast by cheaper gaseous-discharge lamps, especially mercury vapor, sodium vapor, and fluorescent light. After World War II, especially through the 1950s and 1960s, the very highest light intensities were brought to street lighting, not only in traditional downtown areas but also out along emergent commercial strips. City streets were tinted greenish-blue or golden pink, and form-defining shadows were obliterated. Electric signs, even those brightly backlit, carried less visual power owing to the relative lack of encompassing dark background. Spillover light from streetlights prevented silhouetting because it negated dark shadow. By 1970, Americans could claim that, in their cities at least, night had been turned into day.

Lighting the night in ways to best accommodate motor vehicles fit the modernist's penchant for functionality through standardization. In a world that called for more of everything and embraced the new, brighter was better and the newest kinds of light best of all. Unfortunately, functionality required overt simplicity. What were streets for? Planners and engineers decided that streets were for rapid traffic flow. In the nighttime city, functionality hinged on one dimension—automobilty. Street lighting soon brought a profound sameness to nighttime seeing. Yet there was also a promise of exhilaration to be had moving in a fast car through night-time city streets. Night driving was, in a sense, a blend of light and speed—the fundamental attraction of amusement park rides. Only now light and speed were conjoined anew, brought to full but stultifying visual climax. This took place not in special nighttime entertainment zones but in the common spaces of major city streets. Largely gone was the sense of surprise, and even the sense of mystery, which once characterized streets through contrasts of light and dark. Gone, in large measure, was the romance of the night.

Circumstances have changed somewhat in recent decades. The 1980s brought nighttime rediscovery. Planners, developers, architects, and landscape designers began to put more imagination into nighttime illumination. The single-purpose traffic artery of the lighting engineer was, for selected kinds of urban landscape, rethought. Diverse lighting environments conducive to more varied street life were advocated. Many historical lighting precedents, like "white way" lighting, were widely adopted again in downtown regions and small town main street revitalization efforts. The changes expressed a desire to reaccommodate the pedestrian, furnishing the city street not just with night light but also with fixtures that were attractive in the daytime. The look of city streets was rethought, as they were seen not merely as traffic arteries but as diverse, ambient places. Operating also was the desire to bring architecture alive in the night, to capture and put to work the long-neglected sense of nighttime spectacle through varied lighting configurations. Such redirection and resurgence of past practices today allows us all to be vicarious time travelers.

Setting the fantasy of time travel aside, what are we today to make of nighttime lighting? What might we expect in the future? We enjoy a rich heritage of technological innovation and application. A great diversity of night lighting ideas emerged over time, found adoption, and variously thrived or passed away into obscurity. Generally, cost, not just quality of visual effect, has been the prime determinant of a lighting technology's persisting popularity. Can we ever expect to fully escape the tyranny of cost? The rise of the automobile led to an unprecedented rebuilding of streets and roads, as well as unprecedented construction of new thoroughfares, especially freeways, all of which require high levels of light intensity. The American city was geographically reorganized. As cities became more spread out, street lighting needs were amplified geometrically. Not only was a more intense light needed, but it was also needed over much larger street networks over more extensive urban areas. The various electric gaseous discharge lamps that cut lighting costs through power reduction and less servicing were a welcome development. The American night quickly turned greenish-blue or golden pink with little opposition.

Automobility changed the American city both in terms of density and in terms of land use. Increasingly, Americans segregated themselves and their activities in new ways, geographically separating residences and workplaces as never before. Americans also segregated themselves socially as never before, sorting themselves spatially by income and lifestyle. Affluent, mostly white Americans took to the suburbs, where new housing signified social advance, leaving the older parts of cities to minorities and the poor. Social instability wrought tension. Enhanced lighting for security reentered lighting equations starting in the 1960s, especially after civil disorder wracked many of the nation's metropolises. The shadow-busting light of mercury vapor and sodium vapor found favor as both a nighttime guardian and cost reducer. The result was a general heightening of nighttime lighting homogeneity.

Still the call for more diversity in nighttime lighting persisted. And might we not expect it to continue? As the so-called postmodern era in architecture, characterized by history-based example (modernism parodied through embrace of historicism), matures, might not we expect more borrowing from past lighting experience? If so, what would be next? It likely would not mean the reintroduction of old lamp technologies. That would stand as little more than technical stunt. Rather, the search would center on certain prized qualities of night light. Limited reintroduction of gas fixtures, for example, has certainly proven successful: pools of flickering light in dark shadow set a romantic mood at night, especially outdoors in upscale residential neighborhoods. But these are new gas mantles at work, not the old lamps themselves. A past lighting effect has been simulated through up-to-date, cost-efficient fixtures. The future, therefore, might indeed offer a wider palette of different kinds of light once known, with greater contrasts of light and shadow configured in ways not only functional but also aesthetically pleasing.

A new philosophy of night lighting is emerging. Perhaps it will fulfill many of the promises made in years past at the world's fairs. Earlier lighting at the fairs reminds us of how much we lost—or, more correctly, how much we never fully exploited in cities. Each fair, in its own way, made a statement about how future cities might be lit. Maybe only one fair, the 1964 New York World's Fair, correctly predicted. Or it might be more correct to say that this fair represented an already largely fulfilled prophecy. There the master highwayman, Robert Moses, acting as the fair's custodian, approved use of the same mercury vapor fixtures then currently being installed on bridges, along freeways, and as replacement lamps along the major streets of New York City. But what of the pastel colorings of Chicago's 1933 Century of Progress? Or the shimmering light effects of San Francisco's 1915 exposition? What might such nighttime lighting schemes offer us? Already architects have adopted new ways of lighting buildings at night that are at once innovative and substantially reminiscent of floodlighting and light festooning schemes once common in the past. Yet how might whole landscapes be lit in the night for overall visual effect?

Basically, nighttime lighting innovation would seem to hinge on controlling street lighting, directing it to a multiplicity of ends. Have the signs of a new street lighting philosophy appeared? This is the conclusion reported recently in one trade journal. Safety, orientation, security, and aesthetics are all of fundamental concern, wrote Heather Duval in assessing the street lighting program of Columbus, Ohio, in the mid-1990s. *Safety* means the ability of users, both motorists and pedestrians, to reach destinations without "causing inadvertent physical harm to themselves or others."[1] *Orientation* involves visual cuing of location, direction, route, and destination. It means enabling users of the night landscape not only to see where they are going but also to effectively judge their progress. *Security* implies "freedom from deliberate harm or threat by others."[2] *Aesthetics* is, overall, a sense of orderliness and a matter of visual appeal; it involves how lighting can make the night appear satisfying. "Street lights should be compatible with their surroundings during the day, should not contribute to visual clutter, and should help transform a street into an attractive, public place for its users day or night," Duval wrote.[3] Such advocacy represents a start, but, on the face of it, it seems a haltingly feeble step forward, an overly cautious look to the future.

Moving forcefully ahead, however, are the Las Vegas developers. Over $2 billion in new construction was completed in recent years on the Las Vegas Strip alone. New casinos have lent themselves to a new mode of exaggerated theme architecture, fully incorporating the thrill of the amusement park. As a testament to this new architecture stands the 1,149-foot Stratosphere Tower, now the tallest structure west of the Mississippi River. A revolving restaurant sits atop with a spectacular night view, while a daredevil roller coaster spirals around the tower's apex. The New York, New York is an only slightly miniaturized (or so it appears) cloning of the Empire State Building, the Chrysler Building, and a pastiche of other New York City landmark structures, the whole encircled by a "Coney Island Roller Coaster." Upping the ante of streetside "curb appeal," the casino's great sign includes a facsimile of the Statue of Liberty, floodlit in the night with uplifted torch ablaze. In the older section of downtown, Fremont Street, now completely closed to traffic, stands refurbished as a pedestrian mall—"a nave with an 80-foot high cathedral vault of lights stretching five blocks."[4] Light shows are a nightly attraction. "After dark, during hourly shows" reported the New York Times, lights "blink with images of fighter jets buzzing the neighborhood or snowmen dancing to a jingle. On one recent evening, thousands of people craned their necks to watch."[5]

Notes

Introduction: City Lights in the American Imagination

1. Jack Finney, *Time and Again* (New York: Paperback Library, 1970).

2. Ibid., 129.

3. Ibid.

4. "Turning off the Gas in Paris," *Electrical Review* 32 (Sept. 18, 1886): 4, quoted in Carolyn Marvin, "Dazzling the Multitudes: Imagining the Electric Light as a Communications Medium," in Joseph J. Corn, ed., *Imagining Tomorrow: History, Technology, and the American Future* (Cambridge: MIT Press, 1986), 203.

5. Ibid.

6. Quoted in Ben Bova, *The Beauty of Life* (New York: John Wiley and Sons, 1988), 84.

7. Edward Winslow Martin, *The Secrets of the Great City* (Philadelphia: Jones, Brothers, 1868), 44.

8. Ibid., 45.

9. Ibid.

10. "Broadway," *Harper's New Monthly Magazine*, June 1862, 4.

11. David C. Hammack, "Developing for Commercial Culture," in William R. Taylor, ed., *Inventing Times Square: Commerce and Culture at the Crossroads of the World* (Baltimore: Johns Hopkins University Press, 1996), 45.

12. Mark S. Foster, "The Automobile and the City," in David L. Lewis, ed., "The Automobile and American Culture" (special issue), *Michigan Quarterly Review* 19 and 20 (fall 1980 and winter 1981): 464.

13. Ronald J. Horvath, "Machine Space," *Geographical Review* 64 (April 1974): 167–88.

14. For further discussion, see John A. Jakle, "Landscapes Redesigned for the Automobile," in Michael P. Conzen, ed., *The Making of the American Landscape* (Boston: Unwin Hyman, 1990), 293–310; and Edward Relph, *The Modern Urban Landscape* (Baltimore: Johns Hopkins University Press, 1987).

15. Sinclair Lewis, *Babbitt* (1922; New York: Signet Classic, 1950), 178.

16. Ibid.

17. John Dos Passos, *Manhattan Transfer* (New York: Harper, 1925), 216.

18. For an excellent history of Las Vegas and the rise of the Strip, see Alan Hess, *Viva Las Vegas: After Hours Architecture* (San Francisco: Chronicle Books, 1993).

19. Robert Venturi, Denise Scott Brown, and Steven Izenour, *Learning from Las Vegas: The Forgotten Symbolism of Architectural Form* (Cambridge: MIT Press, 1977), 17.

20. Ibid.

21. Ibid, 18.

22. Peter S. Beagle, *I See by My Outfit* (New York: Ballantine Books, 1965), 163.

23. Ibid.

24. Ibid., 171.

25. Tom Wolfe, *The Kandy-Kolored Tangerine-Flake Streamline Baby* (New York: Pocket Books, 1966), 5.

26. Charles Mulford Robinson, *Modern Civic Art, or the City Made Beautiful* (New York: G.P. Putnam and Sons, 1903), 5.

27. F. Laurent Godinez, *Display Window Lighting and the City Beautiful* (New York: Wm. T. Comstock, 1914), 26, 30.

Chapter One: Oil and Gas Lighting

1. Louis Bader, "Gas Illumination in New York City, 1823–1863," Ph.D. diss., New York University, 1970, 33.

2. Harold F. Williamson and Arnold R. Daum, *The American Petroleum Industry: The Age of Illumination, 1859–1899* (Evanston, Ill.: Northwestern University Press, 1959), 29.

3. The luminosity of light has been measured variously over the past two centuries. A candle flame burning under specified conditions was taken early as a standard measure of brightness in gas and then electric lamps. But since the 1930s, the International Commission on Illumination (Commission Internationale de l'Eclairage, or CIE) has worked to establish the candela as the world's standard. A candela is of such magnitude that the luminance of a perfectly radiating "black body" at melting point of platinum is 60 candelas per square centimeter. In the United States this standard is accepted, but the old term *candle,* rather than *candela,* is preferred. Thus candlepower is the light-radiating capacity of a light source in a given direction in terms of the luminosity expressed in candelas or candles. The light (or luminance flux) falling on one square foot of surface one foot away from a point source of one candle power is called a *lumen.* The area of a sphere with a radius of one foot is 4 pi, or 12.57 square feet. It follows, therefore, that a light source having a luminous intensity of one candle in all directions emits 12.57

lumens. When one lumen of light is evenly distributed over a one-square-foot surface, that same area is illuminated to an intensity of one foot-candle. Lamp efficiency is expressed in lumens per watt. The term *brightness* is generally avoided
due to its ambiguous, nontechnical implications. Brightness, in the vernacular
sense, refers to visual sensation and not to photometric quality. One lumen of
white light from an electric incandescent lamp is about 0.004 of a watt. Lighting
engineers evaluate luminance in terms of the unit measure called a *lux*, which is
equal to roughly 18 watts per square foot of energy. One lux approximates one
meter-candle, or one lumen per square meter. Named after Scottish inventor James
Watt, a watt is the energy required to do the work of one English horsepower and
originally evolved as a way of calculating work of steam engines.

4. Henry David Thoreau, *Walden, Or Life in the Woods* (New York: New
American Library, Signet Classic, 1960), 92.

5. Matthew Luckiesh, *Torch of Civilization: The Story of Man's Conquest of
Darkness* (New York: G.P. Putnam's Sons, 1940), 72.

6. Bader, "Gas Illumination," 281.

7. Luckiesh, *Torch of Civilization*, 73.

8. Bader, "Gas Illumination," 218.

9. [Anon.], *A Traveler, Sketches of Highway Life, and Manners in the United
States* (New Haven: privately printed, 1826), 108.

10. Wolgang Schivelbusch, *Disenchanted Night*, trans. Angela Davies (Berkeley: University of California Press, 1988), 97.

11. [Anon.], *A Traveler*, 108.

12. Carl Bridenbaugh, *Cities in Revolt* (New York: Alfred A. Knopf, 1955), 242.

13. Bader, "Gas Illumination," 37.

14. Ibid., 39.

15. Ibid., 238, 249.

16. Williamson and Daum, *The American Petroleum Industry*, 34.

17. M[atthew] Luckiesh, *Artificial Light: Its Influence upon Civilization* (New
York: Century, 1920), 60.

18. "The Year's Progress in Street Lighting," *American City* 34 (Jan. 1926): 42;
"Philadelphia Studies Street-Lighting Problems," *American City* 45 (Jan. 1931):
146.

19. Frederick L. Collins, *Money Town* (New York: G.P. Putnam's Sons, 1946),
183.

20. See Schivelbusch, *Disenchanted Night*, 28.

21. Frederick Moor Binder, "Gas Light," *Pennsylvania History* 22 (Oct. 1955):
361.

22. Ibid., 363.

23. *New York Evening Post*, June 12, 1824, quoted in Bader, "Gas Illumination," 126.

24. *Springfield Republican*, Oct. 13, 1824, quoted ibid.

25. Ibid.

26. Ibid.

27. Frances Trollope, *Domestic Manners of the Americans* (New York: Vintage Books, 1960), 352.

28. Bader, "Gas Illumination," 173.

29. Charles Lockwood, *Manhattan Moves Uptown* (Boston: Houghton Mifflin, 1976), 197.

30. Moses King, *King's Handbook of New York City* (Boston: privately printed, 1893), 51.

31. Luckiesh, *Torch of Civilization*, 108.

32. Lockwood, *Manhattan Moves Uptown*, 197.

33. Bader, "Gas Illumination," 3.

34. Harold L. Platt, *The Electric City: Energy and the Growth of the Chicago Area, 1830–1930* (Chicago: University of Chicago Press, 1991), 14.

35. [Anon.], *A Traveler*, 203.

36. James S. Buckingham, *America: Historical, Statistical, and Descriptive* (London: Fisher, Son, and Co., 1841), 1:221.

37. Nadja Maril, *American Lights: 1840–1940* (West Chester, Pa.: Schiffer, 1989), 38.

38. *New York Mirror*, June 19, 1824, quoted in Bader, "Gas Illumination," 148.

39. *New York Gazette and General Advertiser*, May 20, 1825, quoted ibid., 156.

40. Ibid., 157.

41. See Neil Harris, *Humbug: The Art of P. T. Barnum* (Boston: Little, Brown and Co., 1973), 54.

42. Lloyd Morris, *Incredible New York: High Life and Low Life of the Last Hundred Years* (New York: Random House, 1951), 7.

43. See Gunther Barth, *City People: The Rise of Modern City Culture in Nineteenth-Century America* (New York: Oxford University Press, 1980), 110.

44. Robert Louis Stevenson, "The Lamplighter," in *A Child's Garden of Verses* (New York: Charles Scribner's Sons, 1895), 37.

45. Arthur A. Bright, Jr., *The Electric-Lamp Industry: Technological Change and Economic Development from 1800 to 1947* (New York: Macmillan, 1949), 212.

46. W. Rayner Straus, "Baltimore Snuffs More Gas Lights," *American City* 70 (June 1955): 157.

Chapter Two: Electric Lighting

1. James Cox, *A Century of Light* (New York: Routledge, 1979), 22.

2. Francis B. Crocker, *Electric Lighting* (New York: Van Nostrand, 1896), 9.

3. Cox, *A Century of Light*, 25.

4. P. W. Kingsford, *Electrical Engineering: A History of the Men and the Ideas* (New York: St. Martin's Press, 1969), 121.

5. Paget Higgs, *The Electric Light in Its Practical Application* (London: E. and F. N. Spor, 1879), 200.

6. Professor Silliman, quoted in Preston S. Millar, "Development of Street Lighting Equipment," *Electrical Review* 76 (May 8, 1920): 770.

7. Charles F. Brush, "Some Reminiscences of Early Electric Lighting," *Journal of the Franklin Institute* 206 (July 1928): 11.

8. Quoted in Clarkson W. Weesner, *History of Wabash County, Indiana* (Chicago: Lewis Publishing Co., 1914): 1:311–12.

9. See Dorothea Bump, "Lights Turned on for the First Time in 1880," *Muncie Evening News,* Feb. 15, 1947, quoted in David E. Nye, *Electrifying America: Social Meaning of a New Technology, 1880–1940* (Cambridge: MIT Press, 1990), 3.

10. Arthur A. Bright, Jr., *The Electric-Lamp Industry: Technological Change and Economic Development from 1800 to 1947* (New York: Macmillan, 1949), 31.

11. Ibid., 32.

12. Mark J. Bouman, "City Lights and City Life: A Study of Technology and Urbanity," Ph.D. diss., University of Minnesota, 1984, 272.

13. Harold L. Platt, *The Electric City: Energy and the Growth of the Chicago Area, 1880–1930* (Chicago: University of Chicago Press, 1991), 30.

14. Nye, *Electrifying America, 32.*

15. Platt, *The Electric City, 22–23.*

16. Brush, "Some Reminiscences," 7.

17. Ibid., 9.

18. Nye, *Electrifying America, 5.*

19. Ibid., 31.

20. Joseph W. Wetzler, "The Lighting of Fifth Avenue," *Electrical Engineer* 14 (Nov. 2, 1892): 430.

21. Moses King, *King's Handbook of New York City* (Boston: privately printed, 1893), 51.

22. C. F. Lacombe, "Street Lighting Systems and Fixtures in New York City," *American City* 8 (May 1913): 517.

23. *Chicago Tribune,* April 26, 1878, quoted in John Hogan, *A Spirit Capable: The Story of Commonwealth Edison* (Chicago: Mobium Press, 1986), 12.

24. "The Electric Light Progress in New Orleans, La.," *Electrical Review* 5 (Jan. 10, 1885): 2.

25. Wolgang Schivelbusch, *Disenchanted Night,* trans. Angela Davies (Berkeley: University of California Press, 1988), 125.

26. Bouman, "City Lights and City Life," 275.

27. "Washington," *Electrician and Electrical Engineer* 3 (May 1884): 111.

28. Bouman, "City Life and City Lights," 13.

29. Eddy S. Feldman, *The Art of Street Lighting in Los Angeles* (Los Angeles: Dawson Book Shop, 1972), 29.

30. George B. Catlin, *The Story of Detroit* (Detroit, 1923), 608.

31. Schivelbusch, *Disenchanted Night,* 127.

32. "Electricity in the West," *Electrician and Electrical Engineer* 3 (Feb. 1884): 27.

33. See *Roadway Lighting Handbook* (Washington, D.C.: U.S. Department of Transportation, 1978), 86.

34. R. B. Hussey, "The Luminous Arc Lamp," *General Electric Review* 24 (Aug. 1921): 729.

35. M[atthew] Luckiesh, *Artificial Light: Its Influence on Civilization* (New York: Century, 1920), 159.

36. Schivelbusch, *Disenchanted Night,* 71.

37. Brush, "Some Reminiscences," 7.

38. "Correspondence: New York City and Vicinity," *Electrical Engineer* 8 (July 1889): 322.

39. "The Electrical Light," *Harper's New Monthly Magazine,* Aug. 1870, 358.

40. W. E. Adams, *Our American Cousins* (London: Walter Scott, 1883), 202.

41. Ibid., 202–3.

42. "Correspondence: Chicago," *Electrical Engineer* 10 (July 9, 1890): 49.

43. Rob Kroes, *High Brow Meets Low Brow: American Culture as an Intellectual Concern* (Amsterdam: Free University Press, 1988), 10.

Chapter Three: Maturation of the Lighting Industry

1. Olin J. Ferguson, *Electric Lighting* (New York: McGraw- Hill, 1920), 50.

2. James A. Cox, *A Century of Light* (New York: Benjamin, 1979), 28.

3. Ibid., 26.

4. Ibid., 28.

5. Thomas P. Hughes, *Networks of Power: Electrification in Western Society, 1880–1930* (Baltimore: Johns Hopkins University Press, 1983), 22.

6. W. J. Lampton, "Electricity," in *Thirty Years of New York, 1882–1912* (New York: Press of the New York Edison Co., 1913), 246–47.

7. William Archer, *America To-Day: Observations and Reflections* (London: William Heinemann, 1900).

8. Quoted in *Thirty Years of New York,* 28.

9. Arthur A. Bright, Jr., *The Electric-Lamp Industry: Technological Change and Economic Development from 1800 to 1947* (New York: Macmillan, 1949), 70.

10. Ibid., 71.

11. "Chicago," *Electrical Engineer* 8 (Nov. 1889): 485.

12. *Thirty Years of New York,* 291.

13. W. Parker Chase, *New York: Wonder City* (New York: Wonder City Publishing Co., 1932), 201.

14. Gary Hoover, Alta Campbell, and Patricia J. Spain, eds., *Hoover's Handbook, 1991* (Emeryville, Calif.: Publishing Group West, 1991), 182.

15. "The Edison Incandescent System," *Electrician and Electrical Engineer* 4 (Oct. 1885): 399.

16. Hoover et al., *Hoover's Handbook, 1991,* 178.

17. See Harold L. Platt, *The Electric City: Energy and the Growth of the Chicago Area, 1880–1930* (Chicago: University of Chicago Press, 1991), 68.

18. Hughes, *Networks of Power,* 370–71.

19. Platt, *The Electric City,* xvi.

20. Hoover et al., *Hoover's Handbook, 1991,* 31.

21. Ibid., 75.

22. Bright, *The Electric-Lamp Industry,* 98.

23. "Electric Light and Power," *Electrical Engineer* 7 (Jan. 1888): 38.

24. Cox, *A Century of Light,* 38.

25. Moses King, *King's Handbook of New York City* (Boston: privately printed, 1893), 926.

26. Bright, *The Electric-Lamp Industry,* 151.

27. Ibid.

28. Cox, *A Century of Light,* 214.

29. Matthew Luckiesh, *Torch of Civilization: The Story of Man's Conquest of Darkness* (New York: G.P. Putnam's Sons, 1940), 161.

30. Bright, *The Electric-Lamp Industry,* 134.

31. Ibid., 136.

32. Platt, *The Electric City,* 145–46.

33. *Street Lighting Manual* (New York: Edison Electric Institute, 1969), 6.

34. Ibid., 7.

35. "Gas Street Lights Replaced by Electric Lights," *Electrical Review* 14 (Jan. 18, 1919), 102.

36. Ibid.

37. Albert Schieble, "Electric Street Lighting—3," *Electric Lighting and Illuminating Engineering* 58 (April 15, 1911): 750.

38. M[atthew] Luckiesh, *Artificial Light: Its Influence upon Civilization* (New York: Century, 1920), 162.

Chapter Four: Cutting Costs and Brightening the Night

1. M. J. Francisco, "Municipal Lighting," *Electrical Engineer* 10 (Aug. 27, 1890): 210.

2. David E. Nye, *Electrifying America: Social Meanings of a New Technology, 1880–1940* (Cambridge: MIT Press, 1990), 179.

3. "Municipal Street Lighting Practices Surveyed," *American City* 69 (Jan. 1954): 139.

4. Arthur A. Bright, Jr., *The Electric-Lamp Industry: Technological Change and Economic Development from 1800 to 1947* (New York: Macmillan, 1949), 219.

5. Ibid., 220.

6. Ibid., 224.

7. James A. Cox, *A Century of Light* (New York: Benjamin, 1979), 66.

8. Ibid., 67.

9. Bright, *The Electric-Lamp Industry,* 227.

10. *Street Lighting Manual* (New York: Edison Electric Institute, 1969), 7.

11. Bright, *The Electric-Lamp Industry,* 375.

12. *IES Lighting Handbook* (New York: Illuminating Engineering Society, 1969), 8–29.

13. *Street Lighting Manual,* 113.

14. E. L. Elliott, "Industrial Lighting with Mercury Vapor Lamps," *Electrical Review* 79 (Sept. 10, 1921): 400.

15. "Latest in Mercury-Vapor Street Lamps," *American City* 63 (May 1948): 133.

16. John W. Young, "Street Lamp Development—Mercury, Sodium Vapor and Fluorescent," *American City* 69 (Dec. 1954): 107; John W. Young, "Sizes, Efficiencies, Colors Make Mercury Vapors Continue to Gain," *American City* 73 (Feb. 1958): 130.

17. "Mercury Vapors Continue to Gain," *American City* 83 (April 1968): 136.

18. Bright, *The Electric-Lamp Industry,* 378.

19. "Sodium Light 35 Times Moonlight for San Francisco-Oakland Bridge," *American City* 50 (Sept. 1935): 124.

20. *IES Handbook,* 8–23.

21. Samuel G. Hibben, "Street and Highway Lighting on a Better-Seeing Basis," *American City* 48 (Dec. 1933): 87.

22. Leslie J. Sorenson, "Chicago's Sodium-Lighted Intersections," *American City* 53 (July 1938): 111.

23. John E. Baerwald, *Traffic Engineering Handbook* (Washington, D.C.: Institute of Traffic Engineers, 1965), 581.

24. "A Sodium-Vapor Attack on Street Crime," *Business Week,* May 12, 1973, 173.

25. E. A. Wareham, "Lighting Changes That Save Energy," *American City and County* 96 (Jan. 1981): 45.

26. R. James Claus and Karen E. Claus, *Visual Environment: Sight, Sign and By-law* (Don Mills, Ontario: Collier-Macmillan, 1971), 40.

27. Rudi Stern, *Let There Be Neon* (New York: Harry N. Abrams, 1979), 173.

28. Bright, *The Electrical-Lamp Industry,* 382.

29. Ibid., 384.

30. Cox, *A Century of Light,* 69.

31. Ibid.

32. A. F. Dickerson, "Fluorescent Street Lighting—Why, When, Where," *General Electric Review* 54 (May 1951): 46.

33. *Street Lighting Manual,* 8.

34. John E. Traister, *Principles of Illumination* (Indianapolis: Bobbs-Merrill, 1974), 25.

35. *Street Lighting Manual*, 119.

Chapter Five: Lighting City Streets

1. *Street Lighting Manual* (New York: Edison Electric Institute, 1969), 63.

2. Ibid., 64.

3. T. Commerfield Martin, "Modern American Methods of Street Illumination," *American City* 11 (1914): 398.

4. Ward Harrison, O. F. Haas, and Kirk M. Reid, *Street Lighting Practice* (New York: McGraw-Hill, 1930), 14.

5. Preston S. Millar, "The Lighting of Streets—Part 1," in *Illuminating Engineering Practice* (New York: McGraw-Hill, 1917), 418.

6. *Street Lighting Manual*, 13.

7. Viggo Bech Rambusch, "The Marks Legacy," *Lighting Design and Application* 23 (May 1993): 68.

8. *Street Lighting Manual*, 12.

9. James M. Tien, *Street Lighting Projects* (Washington, D.C.: National Institute of Law Enforcement and Criminal Justice, U.S. Department of Justice, 1979), 256–57.

10. Vincent P. Gallagher, *New Directions in Roadway Lighting* (Washington, D.C.: U.S. Department of Transportation, 1980), 2.

11. I. W. J. M. Van Bommel and J. B. de Boer, *Road Lighting* (Antwerp: Philipas Technical Library, 1980), 97.

12. See *Street Lighting Manual*, 32–33.

13. *Roadway Lighting Handbook* (Washington, D.C.: U.S. Department of Transportation, 1978), 54–55.

14. *Street Lighting Manual*, 11.

15. Ibid., 32.

16. Millar, "The Lighting of Streets—Part 1," 438.

17. Ibid.

18. *Roadway Lighting Handbook*, 118.

19. "Ornamental Street Lighting for an Entire City," *American City* 9 (Dec. 1913): 555.

20. Ward Harrison, "Relighting Indianapolis," *American City* 34 (Feb. 1926): 163–64.

21. R. L. Biesele, Jr., "The Science of Safety Lighting," *American City* 54 (July 1939): 121.

22. F. A. Vaughan, "Some Distinctive Features of Milwaukee's New Street Lighting System," *American City* 15 (Sept. 1916): 303.

23. U. S. Department of Transportation, Federal Highway Administration, *Highway Statistics: Summary to 1985* (Washington, D.C.: 1986), Table MV-201.

24. James A. Cox, *A Century of Light* (New York: Benjamin, 1979), 80.

25. "An Illustrated Chart of Automotive Headlamp History—1892–1941," *Magazine of Light,* Oct. 4, 1941, 24–25.

26. E. J. Edward and H. H. Magdick, "Light Projection: Its Applications," in *Illuminating Engineering Practice* (New York: McGraw-Hill, 1917), 221.

27. W. E. Underwood, "The Trend of Modern Practice in Street Lighting," *Electrical Review* 52 (Nov. 1937): 135.

28. L. A. S. Wood, "Pedestrians Should Be Seen and Not Hurt," *American City* 52 (Nov. 1937): 135.

29. Dudley M. Diggs, "Light, the City's Great Protector," *American City* 48 (1933): 100.

30. L. J. Schrenk, "'Economizing' on Street Lighting in the City of Detroit," *American City* 48 (Feb. 1933): 93.

31. Howell Wright, "Street-Lighting Accomplishment in Cleveland," *American City* 42 (April 1930): 110.

32. Stephen Carr et. al., *City Signs and Lights* (Boston: Boston Redevelopment Authority and U.S. Department of Housing and Urban Development, 1973), 173.

33. Advertisement for Graybar Street Lighting, *American City* 48 (Oct. 1933): 86.

34. James M. Tien, "Lighting's Impact on Crime," *Lighting Design and Application* 9 (Dec. 1979): 21.

35. J. R. Cravath, "War-Time Street Lighting Economics," *American City* 18 (April 1918): 304.

36. "Officially Approved Blackout Luminaire," *American City* 57 (Aug. 1942): 83.

37. A. F. Dickerson, "Blackouts and Dimdowns in the United States," *American City* 57 (Oct. 1942): 89.

38. Advertisement for the Union Metal Manufacturing Company, *American City* 57 (April 1942): 86.

39. W. T. Perry, "A New Way to Control Traffic on Congested City Streets," *American City* 22 (May 1920): 474.

40. G. F. Prideaux, "Street Traffic Control Apparatus," *General Electric Review* 27 (Feb. 1924): 105.

41. "New Traffic Control System on Broadway," *American City* 31 (Oct. 1924): 330.

42. Charles Le Corbusier, *When the Cathedrals Were White,* trans. Francis E. Hyslop, Jr. (New York: McGraw-Hill, 1964), 69.

43. John E. Baerwald, *Traffic Engineering Handbook* (Washington, D.C.: Institute of Traffic Engineers, 1965), 384.

44. Carl E. Egler and D. C. Young, "East Cleveland Speeds Up Traffic with Light," *American City* 36 (May 1927): 602.

Chapter Six: Light as City Celebration

1. Bruce Bassman, "Fireworks: Living Light," *Lighting Design and Application* 19 (Jan. 1989): 6.

2. William F. Mangels, *The Outdoor Amusement Industry* (New York: Vantage Press, 1952), 186.

3. "Fire-Works at New York on the Fourth," *Harper's Weekly*, July 16, 1859, 458.

4. "Special Electrical Illumination for New York's Fourth of July Celebration," *Electrical Review and Western Electrician* 65 (July 11, 1914): 87.

5. *Thirty Years of New York, 1882–1912* (New York: New York Edison Co., 1913), 7.

6. "The Wide-Awake Parade," *Harper's Weekly*, Oct. 13, 1860, 650.

7. "The Electric Light in Election Return Work," *Electrical Engineer* 24 (Nov. 18, 1897): 484.

8. Ibid., 485.

9. Theodore Waters, "Electricity in the Recent Political Campaign," *Electrical Engineer* 22 (Nov. 11, 1896): 490.

10. Ibid., 495.

11. "San Francisco's Path of Gold," *Journal of Electricity, Power and Gas* 37 (Oct. 14, 1916): 17.

12. "Columbus, 1892," *Electrical Engineer* 14 (Oct. 19, 1892): 362.

13. "Hudson-Fulton Illumination," *Electrical Age* 40 (July 1909): 185.

14. "Hudson-Fulton Celebration," *Electrical Age* 40 (Oct. 1909): 275.

15. Ibid., 276.

16. "Festoons of Incandescents on Chicago Business Streets," *Electrical Engineer* 26 (Dec. 22, 1898): 605.

17. Carolyn Marvin, "Dazzling the Multitude: Imagining the Electric Light as a Communications Medium," in Joseph J. Corn, ed., *Imagining Tomorrow* (Cambridge: MIT Press, 1986), 211.

18. "G.A.R. Illuminations at Washington," *Electrical Engineer* 14 (Oct. 5, 1892): 325.

19. "Knight Templar Illuminations at Pittsburg [sic]," *Electrical Engineer* 26 (Oct. 27, 1898): 409.

20. H. W. Jumper, "Electrical Display at the Knights Templar Conclave," *Electrical Age* 33 (Nov. 1904): 322.

21. Arthur Chapman, "Denver, a Typical American City," *World To-Day* 11 (Sept. 1906): 983.

22. E. Leavenworth Elliott, "The New Street Lighting," *American City* 2 (June 1910): 255.

23. F. Laurent Godinez, *Display Window Lighting and the City Beautiful* (New York: Wm. T. Comstock, 1914), 29.

24. "Electric Floats at Portland," *Journal of Electricity, Power, and Gas* 33 (July 18, 1914): 52; "Electrical Illumination at Festival of Mountain and Plain," *Electrical Review and Western Electrician* 61 (Nov. 9, 1912): 894.

25. James D. McCabe, *New York by Sunlight and Gaslight* (Philadelphia: Douglas Brothers, 1882), 292.

26. Ibid., 293–94.

27. "The Peoples' Christmas Tree," *American City* 13 (Dec. 1915): 534.

28. Sarah Comstock, "Ninety Million Lamps for Christmas Trees," *Literary Digest* 16 (Dec. 23, 1933): 28.

29. Ibid.

30. A. L. Lyman, "The Spirit of Yuletide," *East St. Louis Today*, Nov. 1933, 13.

31. "For Electrical Christmas Communities," *Electrical West* 61 (Nov. 1, 1928): 286.

32. Harry K. Trend and Paul H. Hildebrand, "Practical Christmas Lighting," *American City* 63 (Nov. 1948): 131.

33. "Festive Lights for a Famous Street," *American City* 86 (Dec. 1971): 92.

34. Rufus Steele, "The City That Is," *Sunset* 22 (April 1909): 339.

Chapter Seven: Lighting the World's Fairs

1. Edo McCullough, *World's Fair Midways* (New York: Exposition Press, 1966), 14.

2. Ibid., 29.

3. Reid Badger, *The Great American Fair: The World's Columbian Exposition and American Culture* (Chicago: Nelson Hall, 1979), 9.

4. Neil Harris, *Cultural Excursions: Modern Approaches and Cultural Tastes in Modern America* (Chicago: University of Chicago Press, 1990), 120.

5. David E. Nye, *Electrifying America: Social Meanings of a New Technology, 1880–1940* (Cambridge: MIT Press, 1990), 3.

6. Ibid., 36.

7. J. P. Barrett, *Electricity at the Columbian Exposition* (Chicago: R. R. Donnelley and Sons, 1894), 3.

8. John F. Kasson, *Amusing the Millions: Coney Island at the Turn of the Century* (New York: Hill and Wang, 1978), 18.

9. Badger, *The Great American Fair*, 126.

10. Burton Benedict, *The Anthropology of World's Fairs: San Francisco's Panama-Pacific International Exposition* (London: Scholar Press, 1983), 5.

11. Barrett, *Electricity at the Columbian Exposition*, 7.

12. Kasson, *Amusing the Millions*, 21.

13. Quoted in "The Electric Lighting of the White City," *Electrical Engineer* 16 (July 12, 1893): 35.

14. Quoted in Badger, *The Great American Fair*, 127.

15. John J. Ingalls, "Lessons of the Fair," *Cosmopolitan* 16 (Dec. 1893): 141.

16. Ibid.

17. Theodore Dreiser, *Journalism*, vol. 1, *Newspaper Writings, 1882–1895* (Philadelphia: University of Pennsylvania Press, 1988), 126.

18. Ibid., 137.

19. Quoted in Kasson, *Amusing the Millions*, 23.

20. Clara Louise Burnham, *Sweet Clover, A Romance of the White City* (New York: Grosset and Dunlap, 1894), 201.

21. Robert W. Rydell, *All the World's a Fair: Visions of Empire at America's International Expositions, 1876–1916* (Chicago: University of Chicago Press, 1984), 107.

22. Octave Thanet, "The Trans-Mississippi Exposition," *Cosmopolitan* 25 (Oct. 1898): 610.

23. Ibid., 612.

24. Mabel Balcombe, "Illumination of the Grand Court at the Omaha Exposition," *Electrical Engineer* 26 (July 7, 1898): 19.

25. Rydell, *All the World's a Fair*, 131.

26. Julian Hawthorne, "Some Novelties at the Buffalo Fair," *Cosmopolitan* 31 (Sept. 1901), 486.

27. "The Pan-American Exposition," *Scientific American Supplement* 49 (April 21, 1901): 20, 319.

28. Hawthorne, "Some Novelties at the Buffalo Fair," 484.

29. Ibid.

30. W. E. Goldborough, "Electricity at the St. Louis Exposition," *Electrical Age* 32 (April 1904): 191.

31. Ibid., 189.

32. "Illumination at the St. Louis Exposition," *Electrical Age* 32 (June 1904): 397.

33. McCullough, *World's Fair Midways*, 70.

34. Rydell, *All the World's a Fair*, 187.

35. W. D. Ryan, "Illumination of the Panama-Pacific International Exposition," *General Electric Review* 18 (June 1915): 581.

36. M[atthew] Luckiesh, *Artificial Light: Its Influence upon Civilization* (New York: Century, 1920), 308.

37. Alston Rogers, "Light and Color at the San Diego Exposition," *Magazine of Light* 5 (Jan. 1938): 4.

38. John Hogan, *A Spirit Capable: The Story of Commonwealth Edison* (Chicago: Mobium Press, 1986), 182.

39. *Official Guide Book of the Fair* (Chicago: A Century of Progress, 1933), 20.

40. W. D. Ryan, "Illumination of a Century of Progress Exposition, Chicago, 1933," *General Electric Review* 37 (May 1934): 227.

41. *Official Guide Book of the Fair*, 119.

42. "Electrical Building at the Fair Uses a Mile and a Half of Gaseous Tube Lighting," *Lighting* 24 (Aug. 1933): 22.

43. F. A. Orth, "Luminous Tubing at the World's Fair," *Signs of the Times* 80 (Aug. 1933): 10.

44. "Texas Centennial Exposition," *Signs of the Times* 83 (July 1936): 7.

45. Dean M. Warren, "A Lighting Achievement—The Cleveland Great Lakes Exposition," *General Electric Review* 39 (Sept. 1936): 426.

46. "Lighting at the Golden Gate International Exposition," *Signs of the Times* 91 (March 1939): 22.

47. "Exterior Lighting Details at Treasure Island," *Signs of the Times* 93 (April 1939): 7.

48. McCullough, *World's Fair Midways*, 115.

49. Peter Conrad, *The Art of the City: Views and Versions of New York* (New York: Oxford University Press, 1984): 266.

50. Eugene Clute, "General Motors Exhibit at the New York World's Fair," *Lighting and Lamps* 36 (July 1939): 31.

51. Quoted in Rydell, *All the World's a Fair*, 4.

52. Ibid., 152.

Chapter Eight: Night at the Amusement Parks

1. George F. Foster, *New York by Gas-Light and Here and There a Streak of Sunshine* (New York: M. J. Ivers, 1859).

2. Ibid., 5.

3. For Chicago, see Richard Lindberg, *Chicago by Gaslight: A History of Chicago's Netherworld, 1880–1920* (Chicago: Academy Chicago Publishers, 1996).

4. James D. McCabe, *New York by Sunlight and Gaslight: A Word Description of the Great Metropolis* (Philadelphia: Douglas Brothers, 1882), 252.

5. Ibid., 253.

6. Ibid.

7. William Archer, *America To-Day: Observations and Reflections* (London: William Heinemann, 1900), 48.

8. For concise histories of Coney Island, see Oliver Pilat and Jo Ransom, *Sodom by the Sea: An Affectionate History of Coney Island* (Garden City, N.Y.: Garden City Publishing Co., 1943); Robert E. Snow and David E. Wright, "Coney Island: A Case Study in Popular Culture and Technical Change," *Journal of Popular Culture* 9 (spring 1976): 960–75; John F. Kasson, *Amusing the Millions: Coney Island at the Turn of the Century* (New York: Hill and Wang, 1978).

9. James Huneker, *New Cosmopolis: A Book of Images* (New York: Charles Scribner's Sons, 1915), 165.

10. Ibid., 166.

11. Ibid., 169.

12. Maxim Gorky, "Boredom," *Independent* 63 (Aug. 8, 1907): 309.

13. Ibid., 310.

14. Ibid., 311.

15. Snow and Wright, "Coney Island," 986.

16. Kasson, *Amusing the Millions,* 50.

17. For concise histories of the Disney theme parks, see Margaret J. King, "Disneyland and Walt Disney World: Traditional Values in Futuristic Form," *Journal of Popular Culture* 15 (summer 1981): 116–39; Raymond M. Weinstein, "Disneyland and Coney Island: Reflection on the Evolution of the Modern Amusement Park," *Journal of Popular Culture* 26 (summer 1992): 131–64.

18. Weinstein, "Disneyland and Coney Island," 157.

19. King, "Disneyland and Walt Disney World," 121.

20. Weinstein, "Disneyland and Coney Island," 151; For discussion of Disney's Main Street attractions, see Richard V. Francaviglia, "Main Street USA: A Comparison / Contrast of Streetscapes in Disneyland and Walt Disney World," *Journal of Popular Culture* 15 (summer 1981): 141–56.

21. "Epcot Center Pavilions," *Lighting Design and Application* 13 (Oct. 1983): 21.

22. Deyan Sudjic, *The 100 Mile City* (San Diego: Harvest Original, 1992), 213.

23. John R. Haupt, "Illuminations Say Goodnight," *Light Design and Application* 19 (Jan. 1989): 8.

24. Russel B. Nye, "Eight Ways of Looking at an Amusement Park," *Journal of Popular Culture* 15 (summer 1981): 71.

25. Ibid., 72.

26. See Roger Caillois, *Men, Play, and Games* (New York: Free Press of Glencoe, 1961).

27. Kasson, *Amusing the Millions,* 108.

Chapter Nine: Landmarks Floodlit in the Night

1. "The New Art of Lighting," *Lighting Digest* 32 (April 7, 1906): 516.

2. H. E. Butler, "A Short Cut to the Solution of Floodlighting Problems," *General Electric Review* 29 (April 1926): 260.

3. Tyler Stewart Rogers and Alvin L. Powell, "Exterior Illumintion of Buildings," *American Architect* 18 (July 1935): 63.

4. "The Many Uses of Flood Lighting," *Electrical Review* 71 (Sept. 1, 1917): 362.

5. "Illumination of Building Exterior by Projectors," *American City* 13 (Aug. 1915): 108.

6. Rogers and Powell, "Exterior Illumination," 62.

7. Quoted in William Leach, "Introductory Essay," in William R. Taylor, ed., *Inventing Times Square: Commerce and Culture at the Crossroads of the World* (Baltimore: Johns Hopkins University Press, 1996), 234.

8. "Correspondence—New York and Vicinity," *Electrican and Electrical Engineer* 5 (Dec. 1886): 464.

9. M[atthew] Luckiesh, *Artificial Light: Its Influence upon Civilization* (New York: Century, 1920), 302.

10. Barbara Blumberg, "A National Monument Emerges: The Statue as Park and Museum," in *Liberty: A French-American Statue in Art and History* (New York: Harper and Row, 1986), 216.

11. "Flood Lighting of the Federal Capitol," *Electrical Review and Western Electrician* 70 (June 2, 1917): 939.

12. H. D. M. Grier, "Lighting the Landmarks of Manhattan," *Art in America* 45 (spring / summer 1957): 31.

13. "Niagara in the Limelight," *Literary Digest* 85 (May 30, 1925): 23.

14. Orrin E. Dunlap, "Lighting Niagara Falls by Acetylene," *Electrical Engineer* 24 (July 29, 1897): 482.

15. Quoted in W. D. Ryan, "Illumination of Niagara Falls." *General Electric Review* 10 (Feb. 1908): 118.

16. John W. Hammond, quoted in David E. Nye, *Electrifying America: Social Meanings of a New Technology* (Cambridge: MIT Press, 1990), 58.

17. Ibid.

18. J[ohn] A. Spender, *Through English Eyes* (New York: Frederick A. Stokes, 1928), 18.

19. E. J. Edwards and H. H. Magdick, "Light Projection: Its Applications," in *Illuminating Engineering Practice* (New York: McGraw-Hill, 1917), 239.

20. "Lighting the Wrigley Building, Chicago at Night," *Electrical Review* 79 (Sept. 10, 1921): 397.

21. Ibid.

22. Spender, *Through English Eyes,* 34.

23. "Honoring Lindy with a Man-Made Star," *Literary Digest* 106 (Sept. 20, 1930): 37.

24. C. H. Cloudy, "The Lighthouse Service of the United States," *World To-Day* 12 (May 1907): 536.

25. Lauraline Mack, "Business is Brisk in Olesen's Office," *Magazine of Light* 1 (March 1931).

26. "Progress in Airport Lighting," *American City* 46 (April 1932): 130.

27. A. J. Hawkins, "Birmingham's New Municipal Airport Well Lighted," *American City* 45 (Aug. 1931): 144.

28. W. A. Pennow, "Airport Lighting—Why and How—II," *American City* 53 (Dec. 1938): 66

29. H. K. Flint, "Lighting the Detroit City Airport," *American City* 46 (Jan. 1932): 150.

30. "Lighting Runways of Chicago's New Air Field—O'Hare," *American City* 72 (June 1957): 189.

31. Quoted in "Will Baseball Bugs Become Nighthawks?" *Literary Digest* 105 (May 31, 1930): 34.

32. "More Light on Night Baseball," *Literary Digest* 106 (Sept. 27, 1930): 36.

33. "The Dollars and Cents in Floodlighting Sports," *Literary Digest* 110 (July 11, 1931): 33.

34. O. F. Lyman, "Kewanee Provides Night Baseball in Park," *American City* 45 (Sept. 1931): 156.

Chapter Ten: The Great White Way And Electric Sign Art

1. Margaret Knapp, "Introductory Essay," in William R. Taylor, ed., *Inventing Times Square: Commerce and Culture at the Crossroads of the World* (Baltimore: Johns Hopkins University Press, 1996), 45.

2. Brooks Atkinson, *Broadway* (New York: Macmillan, 1974), 11.

3. Ibid., 179.

4. Ibid., 121.

5. Ken Bloom, *Broadway: An Encyclopedic Guide to the History, People and Places of Times Square* (New York: Facts on File, 1991), xvi.

6. Lloyd Morris, *Incredible New York: Highlife and Lowlife of the Last Hundred Years* (New York: Random House, 1951), 262.

7. Stephen Burge Johnson, *The Roof Gardens of Broadway Theatres, 1883–1942* (Ann Arbor, Mich.: UMI Research Press, 1958), 7.

8. Anne O'Hagen, "A Summer Evening in New York," *Muncey's Magazine,* Aug. 1899, 861.

9. Rupert Hughes, *The Real New York* (London: Smart Set Publishing, 1904), 266.

10. Owen Johnson, *Making Money* (New York: Frederick A. Stokes, 1915), 266.

11. James D. McCabe, *New York by Sunlight and Gaslight* (Philadelphia: Douglas Brothers, 1882), 153.

12. William Archer, *America To-Day: Observations and Reflections* (London: William Heinemann, 1900), 39.

13. Hughes, *The Real New York,* 91.

14. Elijah Brown [Alan Raleigh], *The Real American* (London: F. Palmer, 1913; reprinted, New York: Arno Press, 1974), 46–47.

15. Rudi Stern, *Let There Be Neon* (New York: Harry N. Abrams, 1979), 18.

16. Theodore Dreiser, *The Color of a Great City* (New York: Boni and Liveright, 1923), 119.

17. S. N. Holliday, "Through the Years with Electrical Advertising on the Great White Way," *Signs of the Times* 68 (May 1931): 30.

18. John DeWitt Warner, "Advertising Run Mad," *Municipal Affairs* 4 (June 1910): 274.

19. "From Battery to Harlem: Suggestion of the National Sculpture Society," *Municipal Affairs* 3 (Dec. 1894): 632.

20. Archer, *America To-Day*, 25.

21. Hughes, *The Real New York*, 91.

22. John C. Van Dyke, *The New New York: A Commentary on the Place and the People* (New York: Macmillan, 1909), 212–13.

23. Arthur Williams, "Broadway—A Fascinating Electric Sign Picture Gallery," *Signs of the Times* 35 (March 1917): 22.

24. O. J. Gude, "Art and Advertising Joined by Electricity," *Signs of the Times* 31 (March 1913): 3.

25. Pierre Loti, "Impressions of New York," *Century Illustrated Monthly* 85 (Feb. 1913): 5.

26. Ibid., 612.

27. David E. Nye, *Electrifying America: Social Meanings of a New Technology, 1880–1940* (Cambridge: MIT Press, 1990), 53.

28. Rupert Brooke, *Letters from America by Rupert Brooke* (New York: Charles Scribner's Sons, 1910), 30–31.

29. Advertisement for the O. J. Gude Company, *Signs of the Times* 16 (March 1912): 9.

30. "Brilliant Advertising Tower," *Signs of the Times* 12 (June 1910): 20.

31. "Broadway," *Atlantic Monthly*, June 1920, 854.

32. Vachel Lindsay, "A Rhyme about an Electrical Advertising Sign," in Mike Marquessee and Bill Harris, eds., *New York: An Anthology* (Boston: Little, Brown and Co., 1985), 333.

33. Louis Dodge, "The Sidewalks of New York," *Scribner's Magazine*, Nov. 1921, 584.

34. "A New Star on Broadway," *Poster* 20 (Feb. 1929): 7.

35. "High Cost of 'White Way' Electrical Advertising," *Literary Digest* 80 (Feb. 23, 1924): 72.

36. Jill Stone, *Times Square: A Pictorial History* (New York: Macmillan, 1982), 105.

37. "Warner Brothers Brighten Broadway," *Signs of the Times* 63 (Dec. 1929): 38.

38. "The World's Largest Spectacular," *Signs of the Times* 82 (April 1936): 7.

39. "Times Square New Spectacular," *Signs of the Times* 89 (Dec. 1936): 10.

40. Charles Le Corbusier, *When the Cathedrals Were White* (New York: McGraw-Hill, 1964), 102.

41. Francis Marshall, *An Englishman in New York* (London: G. B. Publications, 1949), 8.

42. "Prismatic Letters High-Light Times Square Spectacular," *Signs of the Times* 131 (July 1952): 22.

43. "48-Foot Plane Cruising over Broadway," *Signs of the Times* 142 (Feb. 1956): 46.

44. Mark McCain, "A Mandated Comeback for the Great White Way," *New York Times,* April 9, 1989.

45. Laurence Senelick, "Private Parts in Public Places," in Taylor, ed., *Inventing Times Square,* 332.

46. "The New 'Night Life' of New York," *Literary Digest* 72 (Feb. 25, 1922): 40.

47. Atkinson, *Broadway,* 432.

48. Allan Ashbot, *An American Experience* (Sydney: Alpha Books, 1966), 91.

49. Stone, *Times Square,* 137.

50. Ashbot, *An American Experience,* 91.

51. McCandlish Phillips, *City Notebook: A Reporter's Portrait of a Vanishing New York* (New York: Liveright, 1974), 23.

52. R. James Claus and Karen E. Claus, *Visual Environment: Sight, Sign and By-law* (Don Mills, Ont.: Collier-Macmillan, 1971), 28.

53. Stephen Carr, *City Signs and Light: A Policy* (Cambridge: MIT Press, 1973), 12.

54. Wilmot Lippincott, *Outdoor Advertising* (New York: McGraw-Hill, 1923), 162.

55. Weston Thomas, "What Is a Sign," *Signs of the Times* 63 (Dec. 1929): 13.

56. Harold L. Platt, *The Electric City: Energy and the Growth of the Chicago Area, 1880–1930* (Chicago: University of Chicago Press, 1991), 147.

57. Williams, "Broadway—A Fascinating Electric Sign Picture Gallery," 23.

58. Benjamin Wall, "The Incandescent Lamp in Connection with Electric Signs," *Electrical Review* 51 (Sept. 28, 1907): 519.

59. "High Cost of 'White Way' Electrical Advertising," 72.

60. O. P. Anderson, "Brief Outline of Electric Sign History and Development," *Signs of the Times* 32 (May 1916): 18.

61. Leonard G. Sheppard, "Sign Lighting," in *Illuminating Engineering Practice* (New York: McGraw-Hill, 1917), 535.

62. Anderson, "Brief Outline," 18; W. B. Goudey, "Use of Tungsten," *Signs of the Times* 16 (Sept. 1911): 24.

63. "50 Years of Electric Signs," *Signs of the Times* 143 (May 1956): 22.

64. "The Lengthening List of Slogan Cities," *Signs of the Times* 35 (Feb. 1917): 7.

65. Stern, *Let There Be Neon,* 36.

66. Claus and Claus, *Visual Environment,* 30.

67. F. A. Orth, "Neon, Its Story," *Poster* 19 (May 1928): 25.

68. "Luminous Tube Signs Meeting with Increased Favor," *Signs of the Times* 55 (Oct. 1926): 62.

69. Advertisement for Claude Neon Lights, Inc., *Signs of the Times* 54 (Aug. 1926): 539.

70. C. A. Atherton, "Electrified Advertising—Its Forms and Design, 3," *Journal of Electricity* 56 (Feb. 1925): 244.

71. M[atthew] Luckiesh, *Light and Color in Advertising and Merchandising* (New York: D. Van Nostrand, 1923), 244.

72. G. R. La Wall, "Structural Requirement for Luminous Fascia Displays," *Signs of the Times* 87 (Sept. 1937): 14.

73. John W. Houck, "Toward a Society of Visual Quality," in John W. Houck, ed., *Outdoor Advertising: History and Regulation* (Notre Dame, Ind.: University of Notre Dame Press, 1969), 34.

74. Lippincott, *Outdoor Advertising,* 279.

75. Carolyn Marvin, "Dazzling the Multitude: Imagining the Electric Light as a Communications Medium," in Joseph J. Corn, ed., *Imagining Tomorrow: History, Technology, and the American Future* (Cambridge: MIT Press, 1986), 212.

76. Charles Mulford Robinson, *Modern Civic Art or the City Made Beautiful* (New York: G. P. Putnam's Sons, 1909), 151.

77. John DeWitt Warner, "Advertising Run Mad," *Municipal Affairs* 4 (June 1910): 282.

Chapter Eleven: Lighting America's Main Streets and Commercial Strips

1. William Leach, "Brokers and the New Corporate Industrial Order," in William R. Taylor, ed., *Inventing Times Square: Commerce and Culture at the Crossroads of the World* (Baltimore: Johns Hopkins University Press, 1996), 99–100.

2. David E. Nye, "Social Class and the Electrical Sublime, 1880–1915," in Robert Kroes, ed., *High-Brow Meets Low-Brow* (Amsterdam: Free University Press, 1988), 11.

3. Mark Jansen Bouman, "City Lights and City Life: A Study of Technology and Urbanity," Ph.D. diss., University of Minnesota, 1984, 198.

4. Wolfgang Schivelbusch, *Disenchanted Night: The Industrialization of Light in the Nineteenth Century,* trans. Angela Davies (Berkeley: University of California Press, 1988), 152.

5. Hiram Blauvelt, "More Light, More Profit!" *Chain Store Age* 5 (Sept. 1929): 82.

6. W. D. Ryan, "Intense Street Lighting," *General Electric Review* 23 (May 1920): 362.

7. Ibid.

8. Quoted in Eddy S. Feldman, *The Art of Street Lighting in Los Angeles* (Los Angeles: Dawson's Book Shop, 1972), 31.

9. Quoted in "Ornamental Street Lighting in America," *Electrical Review* 54 (Aug. 20, 1910): 380.

10. C. L. Eshleman, "Modern Street Lighting," *American City* 6 (March 1912): 517.

11. F. Laurent Godinez, *Display Window Lighting and the City Beautiful* (New York: Wm. T. Comstock, 1914), 25–26.

12. Ward Harrison, "The Cleveland Ornamental Lighting System," *American City* 13 (Dec. 1915): 506.

13. Leo Pfeifer, "America's Brightest Streets," *Signs of the Times* (Jan. 1916): 104.

14. Mayo Fester, "The Municipal Outlook in St. Louis," *American City* 2 (March 1910): 104.

15. Bayard W. Mendenhall, "Illumination at Salt Lake City," *Journal of Electricity* 37 (Oct. 7, 1916): 276.

16. Dudley M. Diggs, "Millions for Street Lighting," *American City* 22 (Jan. 1920): 10.

17. Harold C. Platt, *The Electric City: Energy and the Growth of the Chicago Area, 1880–1930* (Chicago: University of Chicago Press, 1991), 260.

18. Advertisement for the General Electric Company, *American City* 9 (Dec. 1913): 44.

19. E. M. Jenison, "Fond du Lac's Great White Way," *American City* 7 (Sept. 1912): 248.

20. Fred De Land, "The Up-Building of the Small Town," *Electrical Engineer* 21 (Jan. 22, 1896): 101.

21. Sinclair Lewis, *Main Street* (New York: Harcourt Brace Jovanovich, 1961), 400.

22. David E. Nye, *Electrifying America: Social Meanings of a New Technology, 1880–1940* (Cambridge: MIT Press, 1990), 54.

23. See Leonard S. Marcus, *The American Store Window* (New York: Watson-Guptill, 1978), 12–19.

24. William Leach, "Introductory Essay," in Taylor, ed., *Inventing Times Square*, 235.

25. "Turning Off the Gas in Paris," *Electrical Review* 20 (Sept. 18, 1886): 4.

26. "The Electric Light in Stores," *Electricity* 1 (Dec. 30, 1891): 300.

27. S. L. Ruzow, "What Well-Lighted Windows Mean to the Chain Store," *Chain Store Age* 2 (June 1926): 452.

28. "High-Intensity Illumination of Power Company Windows," *Journal of Electricity* 55 (Dec. 15, 1925): 27.

29. Quoted in William Nelson Taft, *The Handbook of Window Display* (New York: McGraw-Hill, 1926), 1.

30. Virginia Roehl, "New York Displays Combine Beauty with Selling Power," *Display World* 56 (Jan. 1950): 41.

31. Viginia Roehl, "Magic Christmas Settings in Manhattan," *Display World* 58 (Jan. 1951): 40.

32. LeRoy Jacks, "Electrical Advertising Opportunities in the Small Cities," *Signs of the Times* 23 (Nov. 1, 1913): 10.

33. J. B. Priestley, *Midnight on the Desert* (New York: Harper and Brothers, 1937): 87.

34. See Richard Mattson, "Store Front Remodeling on Main Street," *Journal of Cultural Geography* 3 (spring / summer 1983): 41–55.

35. Chester H. Liebs, *Main Street to Miracle Mile: American Roadside Architecture* (Baltimore: Johns Hopkins University Press, 1995), 40.

36. Ibid., 42.

37. P. Schuyler Van Bloem, "Architecture versus Electric Signs," *Pencil Points* 10 (April 1929): 265.

38. John B. Jackson, "Other-Directed Houses," *Landscape* 6 (winter 1956–57): 32.

39. Richard Horwitz, *The Strip: The American Place* (Lincoln: University of Nebraska Press, 1985), 9.

40. Liebs, *Main Street to Miracle Mile*, 61.

41. Tom Wolfe, *The Kandy-Kolored, Tangerine-Flake Streamline Baby* (New York: Farrar, Straus and Giroux, 1965), 82.

42. Liebs, *Main Street to Miracle Mile*, 64.

43. Thomas Hine, *Populuxe* (New York: Alfred A. Knopf, 1987), 5.

44. Alan Hess, *Googie: Fifties Coffee Shop Architecture* (San Francisco: Chronicle Books, 1985), 43.

45. Stephen A. Kurtz, *Wasteland: Building the American Dream* (New York: Praeger, 1973), 10.

46. Jackson, "Other-Directed Houses," 31.

47. Ibid., 32.

48. Ibid., 30.

49. "New Extravaganzas: Two Huge Scintillations Join Spectaculars of Las Vegas," *Signs of the Times* 160 (Feb. 1962): 37.

50. See Charles F. Barnard, *The Magic Sign: The Electric Art / Architecture of Las Vegas* (Cincinnati: ST Publications, 1993).

51. Kurtz, *Wasteland*, 12.

52. See: Alan Hess, *Viva Las Vegas: After-Hours Architecture* (San Francisco: Chronicle Books, 1993).

53. Wolfe, *Kandy-Kolored*, 7.

54. Brian O'Doherty, "Highway to Las Vegas," *Art in America* 60 (Jan. / Feb. 1972): 84.

55. "Spectacular's Specifications," *Signs of the Times* 150 (Sept. 1958): 84.

56. Wolfe, *Kandy-Kolored*, 9–10.

57. Hess, *Viva Las Vegas*, 22.

58. Albert Elsen, "Night Lights: Neon Makes Las Vegas a Special Place," *Art News* 82 (Nov. 1983): 113.

59. John Pastier, "The Architecture of Escapism: Disney World and Las Vegas," *AIA Journal* 67 (Dec. 1978): 31.

60. Robert Venturi, Denise Scott Brown, and Steven Izenour, "Learning from Las Vegas," (excerpted from *Learning from Las Vegas* [Cambridge: MIT Press, 1972]) in William H. Ittelson, ed., *Environment and Cognition* (New York: Seminar Press, 1973), 103.

Epilogue

1. Heather Duval, "Lighting Columbus: The City of Columbus Sets Its Sights on 20 / 20 in 2020," *Landscape Architect and Specifier News* 13 (April 1997): 38.

2. Ibid.

3. Ibid.

4. Verne G. Kopytoff, "A New, Dazzling Las Vegas Downtown," *New York Times,* Jan. 28, 1996.

5. Ibid.

Index